AF790721

Precise Dimensions

A history of units from 1791–2018

Precise Dimensions

A history of units from 1791–2018

Malcolm Cooper
The History of Physics Group of the IOP

Jim Grozier
The History of Physics Group of the IOP

IOP Publishing, Bristol, UK

© IOP Publishing Ltd 2017

All rights reserved. No part of this publication may be reproduced, stored in a retrieval system or transmitted in any form or by any means, electronic, mechanical, photocopying, recording or otherwise, without the prior permission of the publisher, or as expressly permitted by law or under terms agreed with the appropriate rights organization. Multiple copying is permitted in accordance with the terms of licences issued by the Copyright Licensing Agency, the Copyright Clearance Centre and other reproduction rights organisations.

Permission to make use of IOP Publishing content other than as set out above may be sought at permissions@iop.org.

Malcolm Cooper and Jim Grozier have asserted their right to be identified as the editors of this work in accordance with sections 77 and 78 of the Copyright, Designs and Patents Act 1988.

Media content for this book is available from Book information at https://doi.org/10.1088/978-0-7503-1487-9.

ISBN 978-0-7503-1487-9 (ebook)
ISBN 978-0-7503-1485-5 (print)
ISBN 978-0-7503-1486-2 (mobi)

DOI 10.1088/978-0-7503-1487-9

Version: 20171101

IOP Expanding Physics
ISSN 2053-2563 (online)
ISSN 2054-7315 (print)

British Library Cataloguing-in-Publication Data: A catalogue record for this book is available from the British Library.

Published by IOP Publishing, wholly owned by The Institute of Physics, London

IOP Publishing, Temple Circus, Temple Way, Bristol, BS1 6HG, UK

US Office: IOP Publishing, Inc., 190 North Independence Mall West, Suite 601, Philadelphia, PA 19106, USA

This book is dedicated to the memory of Dr Bryan Peter Kibble (1938–2016) He was an excellent physicist who spent most of his life working in the area of precision measurements and made many major contributions to this field. He measured the high-field gyromagnetic ratio of the proton, which led to the invention of the watt balance, now renamed the Kibble balance in his honour.

He also spent time measuring the unit of capacitance, the Farad, in terms of the base units of the SI using a device called a calculable capacitor. This process is referred to as 'realising' the unit of capacitance and by a conceptually simple extension to the process it was also possible to realise the SI ohm. This work introduced him to ac coaxial bridges, the beauty of which stayed with him for all his life and inspired him to write two books: Coaxial AC Bridges *with G H Rayner and* Coaxial Electrical Circuits for Interference-free Measurements *with Shakil Awan and Jürgen Schurr. He continued to work in this field after his retirement and played a major part in elucidating the present understanding of the ac Quantum Hall Effect.*

He was best known for his invention of the moving coil watt (or Kibble) balance and I had the privilege of working with him on this almost from its inception. This discovery, and the subsequent global effects resulting from the invention, is the subject of a talk that he gave in 2016. The transcript of this talk is reproduced in this book, and editing has been kept to a minimum to allow his style and sense of humour to be retained in the text. I hope that it will allow you to make the journey with him from his initial invention to its consequence—the redefinition of the kilogram, which is now less than two years away. Everyone who knew him wished that he could have been there to see it.

Ian Robinson, June 2017

Contents

The Editors

Malcolm Cooper is a former broadcast television engineer now retired. His upbringing in the somewhat authoritarian style of engineering studies drove him to pursue an honours degree in physics with the Open University. During this time the IOP opened its doors to OU students and he joined as a student member in 1987, opting for the History of Physics Group, later attaining full corporate membership. He has served on the group's committee in several capacities for many years, including briefly as chairman and secretary but his chief contribution has been as editor of the group's newsletter from 2004 to date. Although sometimes now described as more like a journal, he considers its less formal style a very positive attribute and likes to think it has been instrumental in doubling the group membership over that period.

Jim Grozier is a former railway telecommunications engineer and experimental particle physicist, now working as a lab demonstrator at University College London. His research interests include the philosophy of measurement in the physical sciences, and popular (mis)conceptions of special relativity. He is a member of the Committee of the History of Physics Group, and has published numerous articles in the Group's newsletter. He is the author of *Made In Hungary*, the official history of the International Association of Physics Students, and of a web-based *History of Early High Energy Physics Research at UCL*. His article on bubble chamber scanners at UCL in the 1960s was published in the British Society for the History of Science's magazine, *Viewpoint*, in October 2015.

Preface

During the International Conference on the History of Physics in Cambridge, UK in September 2014, Terry Quinn (a former Director of the *Bureau Internationale des Poids et Mesures*) suggested to Ted Davis (then Chair of the History of Physics Group) that the Group should organise a meeting on the history of units.

The History of Physics Group holds a number of meetings every year, often marking anniversaries of milestones in the history of physics, or the careers of prominent physicists. What Terry suggested, however, was subtly different: in this case, one of the 'milestones' was actually in the future! In 2018 the 26th CGPM (*Conférence générale des poids et mesures*)—the governing body of international metrology—is expected to redefine four of the SI base units in terms of fundamental constants, bringing to an end the need for physical standards such as the International Prototype Kilogram. It was therefore felt that a meeting on the theme of 'the history of units' would be entirely appropriate, and preparations began for the meeting. In order to sharpen the focus a little, a start date of 1791—when the base units of the metric system were first defined—was chosen.

The meeting, 'A History of Units from 1791 to 2018', took place at the National Physical Laboratory, Teddington, UK on 17 March 2016 and was organised by Jim Grozier and Terry Quinn. However, as this redefining represents such a profound change for the physics community and indeed the scientific world, following the meeting, it was suggested by Charlotte Ashbrooke, of the IOPP, that interest in this topic would easily merit the publication of an e-book.

Acknowledgments

We should like to thank all our authors who have worked tirelessly on this project, which would not have been possible without their tremendous support. We should also like to thank the UK Institute of Physics History of Physics Group for their inspirational meeting on the history of units held at the National Physical Laboratory. Many thanks go to Charlotte Ashbrooke of the IOPP for her constant encouragement and to Dan Heatley for his continued assistance and for so tolerantly leading us through the intricacies of copyright permissions.

Finally we should like to extend our warmest thanks to Mrs Anne Kibble for allowing us to include the last public lecture by her husband, and lending her support at that most difficult time.

Author biographies

Jim Grozier - see in The Editors

Sophie Osiecki

Sophie Osiecki completed a BSc in the History and Philosophy of Science at University College London in 2014. She then went on to complete an MPhil in the History, Philosophy and Sociology of Science, Technology and Medicine at the University of Cambridge, graduating in 2016. She is now eagerly looking for an opportunity to start a PhD program in 2018.

Rory McEvoy

Rory McEvoy is Curator of Horology at the Royal Observatory, Greenwich. Before joining the Museum in 2010, he worked for three years at Bonhams auctioneers as a horological specialist after a decade of working at the bench conserving and restoring horological instruments. Current research and writing interests include: the history of development of timekeeping instrumentation for the observatory and laboratory, the life and work of George Graham FRS (c 1693–1751), the 20th century story of time derivation and distribution and traditional methods of manufacturing clocks and watches.

Edward Davis

Edward A Davis holds the positions of Emeritus Professor at the University of Leicester and Distinguished Research Fellow in the Department of Materials Science and Metallurgy at the University of Cambridge. Previously he undertook research at the Cavendish Laboratory where he co-authored a book with Nobel Laureate Professor Sir Nevill Mott on *Electronic Processes in Non-Crystalline Materials*. In 1980 he was offered a Chair of Physics at the University of Leicester where he served as Head of Department and Dean of Science. He is the Coordinating Editor and Letters Editor of the *Philosophical Magazine*—a condensed matter physics and materials science journal first published in 1798. Professor Davis has recently served as Chairman of the Institute of Physics History of Physics Group. He has written several papers and books on the history of science, including four volumes entitled *Science in the Making* and *J J Thomson: The Discovery of the Electron*, co-authored with Isobel Falconer. He is a Fellow of both the Institute of Physics and the American Physical Society.

Since retirement Professor Davis has been working with the present Lord Rayleigh on preserving the laboratories used by his forebears—the third and fourth Baron Rayleighs.

Graham Machin

Graham Machin has published almost 200 papers and articles about different aspects of temperature measurement and given numerous invited lectures on the topic around the world. He is a Fellow of both the Institute of Physics and the Institute of Measurement and Control, CPhys and CEng. He currently holds visiting professorships at the University of Valladolid (UVa), Spain (Temperature Standards), University of South Wales (Clinical Thermal Imaging) and the University of Strathclyde (Applied thermometry in harsh environments). He was a contributing editor to the two-volume work *Radiometric Temperature Measurement* published by Elsevier in 2010. In October 2012 he was awarded the InstMC Callendar Medal award for improvements to the state of the art in temperature measurement, and in June 2015 a DSc from the University of Birmingham for a thesis entitled *Improvements in Temperature Measurement*. He has been a visiting researcher to institutes in Japan (twice) and China (three times). In 2017 Graham was awarded a Chinese Academy of Sciences Fellowship, in recognition of his world-leading position in and decadal contributions to the science of thermometry. His current research interests span all forms of thermometry, contact, non-contact, thermal imaging, primary and applied. In addition he has research interests as diverse as diabetes, wound management and nuclear decommissioning.

Hasok Chang

Hasok Chang is Hans Rausing Professor of History and Philosophy of Science at the University of Cambridge. Previously he taught for 15 years at University College London, after receiving his PhD in Philosophy at Stanford University following an undergraduate degree at the California Institute of Technology. He is the author of *Is Water* H_2O*? Evidence, Realism and Pluralism* (Springer, 2012), winner of the 2013 Fernando Gil International Prize, and *Inventing Temperature: Measurement and Scientific Progress* (Oxford University Press, 2004), joint winner of the 2006 Lakatos Award. He is also co-editor (with Catherine Jackson) of *An Element of Controversy: The Life of Chlorine in Science, Medicine, Technology and War* (British Society for the History of Science, 2007), a collection of original work by undergraduate students at University College London. He is a co-founder of the Society for Philosophy of Science in Practice (SPSP), and the International Committee for Integrated History and Philosophy of Science. He has recently been the President of the British Society for the History of Science.

Juris Meija

Juris Meija is a research officer at the National Research Council Canada working in the area of chemical metrology and certified reference material development. His expertise lies in theoretical analytical chemistry, isotope ratio measurements, and data analysis. Since 2014 he has served as the Chair of the IUPAC Commission on Isotopic Abundances and Atomic Weights and since 2012 as Titular Member on the

IUPAC Interdivisional Committee on Terminology, Nomenclature and Symbols. He is also IUPAC delegate to the Joint Committee for Guides in Metrology, Working Group 1: Guide to the Expression of Uncertainty in Measurement (GUM), member of the United States Pharmacopeia Expert Panel on Statistics and the Statistics and Uncertainty working group of the Regional Metrology Organization for the Americas (SIM), and has represented Canada in the Consultative Committee for Amount of Substance: metrology in chemistry of the International Committee for Weights and Measures. Juris has published 70+ peer-reviewed publications and 50+ op-ed science articles. He is an avid coder, column editor and member of the international advisory board of the journal *Analytical and Bioanalytical Chemistry* (Springer Nature) and has been actively involved in many recent international activities such as the redefinition of the mole and naming of the new chemical elements.

Teresa Goodman

Teresa Goodman works in NPL's Earth Observation, Climate and Optical (ECO) Group, where she leads work on the realisation, maintenance and dissemination of optical radiation scales and standards and associated R&D to improve these where needed. Teresa has more than 30 years' experience at NPL in optical radiation measurements, and is internationally recognised for her expertise in photometry, mesopic photometry, spectroradiometry, radiometry, spectrophotometry, colorimetry and appearance metrology. She has been responsible for many research projects, ranging from the first radiometric realisation of the candela at NPL, through establishment of a completely new scale for spectral total flux (a World first), to the development of a practical system for mesopic photometry and its subsequent refinement and adoption by the CIE (the International Commission on Illumination). She has also played a leading role in 'sensory metrology' research at NPL and led a highly-successful EU project in this area, on measurement of naturalness (MONAT). She was awarded the Society of Light and Lighting Walsh-Weston Award for lighting research in 2008 and the CIE de Boer Gold Pin in 2015.

Sally Riordan

Sally Riordan is a secondary-school physics teacher in Cambridgeshire. She is an affiliated research scholar at the Department of History and Philosophy of Science at the University of Cambridge, where she helps students to improve their essay-writing skills. Her publications include *The Objectivity of Scientific Measures* in *Studies in History and Philosophy of Science* and *How Experiments Begin: Defining the Kilogram by the Planck Constant*, in *Journal for General Philosophy of Science*. She is currently writing a history and philosophy of the kilogram, to be published with the University of Pittsburgh Press.

Richard Davis

Richard Davis joined the International Bureau of Weights and Measures (BIPM) in 1990 following 18 years at the National Institute of Standards and Technology (NIST, USA). He began at NIST, then known as the National Bureau of Standards (NBS), as a post-doctoral fellow in the electrical standards group. Later, he had technical responsibility for dissemination of the unit of mass from the national prototype of the kilogram. At the BIPM, he worked in the mass department until he retired in 2010 as department head. He continues as a consultant to the BIPM Director. Richard is a Fellow of the American Physical Society and an affiliate member of the Institute of Physics.

Introduction

'A unit of measurement is a definite magnitude of a quantity, defined and adopted by convention or by law.'

Vocabulaire International de Métrologie published by the JCGM

A fair definition but what a wealth of questions, scope and history lie in those two little words 'defined and adopted'!

The practical and economic advantages of standards in weights and measures have been clear from ancient times but the challenges of such standardisation, even within a single social group, were enormous. With the coming of improving travel and international movement those challenges became greater still and it is only since the concept of metrology was born that real progress could begin.

It is difficult, of course, to specify such beginnings with any precision but it may be said that two events underpin the inception of this book and the conference which preceded it. The first was the drafting of a report, in 1791, on the reformation of units for the consideration of the French Assembly—the metric system, where each unit of measurement would be defined in relation to fixed, permanent features of the natural world. The second was the decision in 2011 to redefine four of the SI base units in terms of fundamental constants, to be effected in 2018.

227 years of history just waiting to be explored!

This slim volume, based on the meeting, does not attempt to be a rigorous history of the subject, but offers what we hope is an intriguing glimpse into the sometimes clear-sighted and sometimes *ad hoc* development, over the years, of six of the seven SI base units.

In the light of the talk by Professor Edward Davis on Lord Rayleigh's measurement of the ohm, the meeting organisers felt one on the history of the ampere could be omitted. However, we took the opportunity to include chapters on the two remaining SI base units, the mole and the candela, which were also not covered at the meeting, but nevertheless have fascinating histories that are seldom encountered in the literature, with articles from Dr Juris Meija and Dr Teresa Goodman.

Glimpses in brief

Chapter 1

Part 1: The Making of the Metre

Jim Grozier

How long is a piece of string? Or even a platinum bar? In this case—a metre; or 100 cm or 10 dm or 1000 mm. Or, as it was declared in 1791 in the French Assembly, 'one ten-millionth of the meridian arc from the North Pole to the equator'. But why 10^{-7}? Was it because the second could be defined as the half period of a one metre pendulum (uncertain) or was it because the metre was a practical measure roughly similar to the English yard (unlikely)?

Whatever the reason, it was clear to the gentlemen of the *Academie des Sciences* that an attempt must be made to measure the meridian arc. It was equally clear that this distance could not be measured directly, but that a portion of it could be, and the whole assessed from those measurements.Two French astronomers, Pierre-Françis-André Méchain and Jean-Baptiste-Joseph Delambre, were charged with this monumental task and in 1792 the intrepid pair set off armed with rulers, telescopes and great courage.

Jim Grozier gives a rare account of their trials and tribulations and concludes with an unexpected but perhaps not altogether surprising outcome.

Part 2: The Metre Convention and the BIPM

Sophie Osiecki and Jim Grozier

Thus the standard metre—*metre des archives*—was enshrined in a platinum bar which could be used to produce secondary standards and which would be stable for many years. Unfortunately, it was not so. In this brief account of the 'Metre Convention', Sophie Osiecki and Jim Grozier reveal the shortcomings of the extant physical standards and the scientific community's attempt through international cooperation to solve the many problems which had arisen.

But then the 'politicians' moved in to create their gentle mayhem. However, amazingly, despite this, consensus was finally achieved as new standards were established — the International Prototype Metre and the International Prototype Kilogram.

Chapter 2

From notion to precision: the SI second

Rory McEvoy

The concept of time has yielded fertile ground for the writers of fiction for many years, and in particular, that of travel through time. This is not surprising as most of us at one time or another have thought 'what if…?' or even 'if only…!'.

From HG Wells' classic *The Time Machine* to today's more sophisticated offerings, the *measurement* of time is often taken for granted. And a quick look at the covers of books dealing with this dreamy subject suggests that many of the graphic artists are still firmly wedded to a slice of horological history frequently featuring—as they do—the mechanical clock with its dials, springs and gears.

Clock makers had shown extraordinary mechanical ingenuity in constructing devices to mimic the planetary motions but the better they became the more was revealed of the irregularities of that master timekeeper—the Earth. Thus the conviction of an independent time standard gained strength, weathering a backlash from the astronomers, to finally being realised in the first atomic clocks of the mid-20th century.

Rory McEvoy takes us from the primitive through mechanics, electronics and 'atomics' always in pursuit of that seemingly elusive quantity—the precise SI Second.

Chapter 3

Lord Rayleigh's determination of the ohm

Edward Davis

It is arguable which of the currently accepted base and derived SI units have attracted the most controversy and debate—mass, for sure, has a very chequered history, length and time perhaps less so, but electrical units must be a favourite in this contest.

Imagine a dusty room, in the early 1860s, weak sunlight finding its way in to illuminate the bewhiskered features of James Clerk Maxwell, William Thomson and others on the British Association's committee for standards and units. They speak of the volt as a unit of resistance—and the ohma—a unit of emf—and something called the Galvat.

Well, maybe it wasn't quite like that, but it is clear that electrical units and standards in the mid-19th Century were in a state of flux and considerable uncertainty. Even multiple and submultiple prefixes added to the confusion—where kilo could mean 10^{-3} or 10^{+3}. Something had to be done! And something was done—the next couple of decades saw the beginnings of consistent units and standards—especially of standards, which were eagerly sought by those involved in the emerging telecommunications field. For example, by the mid-1860s the BA had produced a number of standard resistors, although as measurement techniques improved it became clear that 'standard' was rather more variable than it should have been!

Certainly Lord Rayleigh suspected so, which, in the early 1880s, led him to carry out the experiments to test the matter, and forms the basis for our third chapter 'Lord Rayleigh's determination of the ohm' by Ted Davis.

Chapters 4 and 5

Temperature scales: past, present and future: 1700–2050
Graham Machin

Kelvin's Absolute Temperature and Its Measurement
Hasok Chang

It has been said that popular weather forecasters should quote low temperatures in Celsius and high temperatures in Fahrenheit, thus making them sound more extreme! Be that as it may, European weather reports use degrees Celsius—with the odd Fahrenheit equivalent sneaking in here and there. It has taken many years for this standard to be adopted and Fahrenheit is still used extensively elsewhere in the media, especially in the US.

The scientific world has, of course, been using degrees Kelvin for over half a century, but it was also the industrial world which provoked the need for standardisation and reliability in temperature measurement. The fascinating history of the development of these scales, in this chapter by Graham Machin, charts the way from the early 18th Century to, rather intriguingly, the mid-21st.

But temperature scales and concepts were dogged for many years by their arbitrariness—there being little real meaning attached to them until the work of William Thomson and others on thermodynamics. This conceptual development was crucial to the progress of physics and is the reason we include a separate chapter focusing on this serious problem, in Hasok Chang's 'Kelvin's Absolute Temperature and its Measurement'.

When Thomson fashioned his concepts of 'absolute' temperature, his main concern was to make the concept of temperature abstract, i.e., independent of the properties of any particular thermometric substances. He tried out a succession of definitions based on the thermodynamics of ideal heat engines. Initially he attempted to define temperature in terms of the amount of heat difference associated with the production of a unit amount of work in a Carnot engine. Later, he conceived absolute temperature so that the ratio of two temperatures is the same as the ratio of quantities of heat taken in and given out at those temperatures in a Carnot cycle.

But there were difficulties with using such definitions for experimental work, since it was not possible even to approximate an ideal Carnot engine in reality. More generally, it is not trivial to connect an entirely abstract concept with concrete situations in order to make physical measurements possible. Thomson pursued this problem by a process of 'operationalization'—to relate theoretical temperature to the physical world of operations.

Chapter 6

A brief history of the unit of chemical amount

Juris Meija

Gold—that soft, heavy metal with its beguiling lustre—held the allure of riches which drove the alchemists in their quest to conjure it from base metals. And it may be argued that as these ancient sorcerers were the precursors of the modern chemist, so the early assayers in their quest for its purity gave birth to analytical chemistry.

Thus begins Juris Meija with these ideas in his chapter on the history of the mole. But the comparative methods of Robert Boyle and others in the 17th century espoused the less glamourous materials of acids and alkalis, which led to thinking about 'quantity' and 'measurement' of substances.

Another 100 years or so were to pass before Lavoisier laid the foundations of modern chemistry and set its feet on the path to discovering the discrete nature of substances—and that smallest of masses, the molecule, from which, of course, was derived the name 'mole'.

Chapter 7

The history of the SI unit of light, the candela

Teresa Goodman

Who has not gazed at the rows of delicious looking fruit and vegetables laid out temptingly on the supermarket trays, brilliant in their perfection, and wondered, are they really that good? Or at the meat basking in just the right colour temperature of illumination? The purveyors well know the benefits of lighting to improve sales—a point which Teresa Goodman hints at in her opening paragraphs on the history of the candela.

And this is not a trivial point; it is well known that the visual perception of an object involves not just the object and its illumination but also the way the eye and brain interpret these sensations. Of all the quantities and units being discussed in this book, none are so intimately bound up with the human observer, and present such conceptual difficulties and debate, as does this unit.

Early attempts at setting a standard for light—perhaps drawing upon the ideas for mass and length standards—used candles or other flame sources, but soon ran into difficulties as, unlike the kilogram and the metre, there were many variables involved—intensity, colour (or more precisely, spectral distribution) and, its effect on the eye. Indeed, the measurement of optical radiation in terms of its visual effect warrants its own particular name of 'photometry' and even radiometry—measurement in terms of optical energy or power—still has connections with photometry, as is revealed in this history of the candela.

Chapter 8

The Story of Mass Standards 1791–2018

Jim Grozier, Sally Riordan and Richard Davis

Even today one suspects that, in Britain, at least, many look wistfully over their shoulders at the passing of the pound as a measure of weight (lb), but most of those many remain unaware that this measure has a rich history dating back many hundreds of years.

The authors, in 'The story of mass standards' trace this convoluted history, wrestling with Troy, Avoirdupois, copper tubes and brass weights to embracing the concept of mass and ironically finally defining the pound in terms of the kilogram.

The chapter focusses on two examples, viz. the British and the French systems, and by comparison with the situation in France the British seems pretty straightforward. In pre-revolutionary France there were around a quarter of a million different weights and measures and it was probably this state of utter confusion that propelled the drive for standardisation immediately following the re-instatement of law and order in 1791.

Chapter 9

Mass from Energy—a Unit for a Quantum World—(The Kibble Balance)

Bryan Kibble

Soon after the meeting took place, we received the very sad news that Bryan Kibble had passed away. The NPL talk was his last.

It was, therefore, not possible to present a written account of his talk, and while we would normally not consider a transcript as very suited for inclusion in a book, we felt that this would be an exceptional opportunity to 'hear' the man himself give a personal account of the genesis of what is now known as 'The Kibble Balance'.

Malcolm Cooper and Jim Grozier

IOP Publishing

Precise Dimensions
A history of units from 1791-2018
Malcolm Cooper and Jim Grozier

Chapter 1

The metre and the metric system

It is no coincidence that the word ‘metre’ derives from the Greek *metron* (μετρέω) meaning to measure, count or compare, nor that the precursor to the modern day International System of Units (SI) was called the *Metric* system. Length has always been regarded as ‘first among equals’ in any fundamental system of measurement. In this chapter we look at the origins of the metre and its rôle in creating the world’s first international scientific institute, the BIPM.

Part 1: The making of the metre

Jim Grozier

1.1 Introduction

The metre was defined in 1791 as the new French unit of length. However, it was to be another 8 years before it could be calibrated against the unit it replaced, and metre rules manufactured. The story of that period has been well told by Ken Alder in his book *The Measure of All Things*, and also by Charles Gillispie in *Science and Polity in France: The Revolutionary and Napoleonic Years*. In this section I will summarise the methods used to find the metre, as covered by Alder, but also—making use of Méchain and Delambre’s comprehensive report *Base du système métrique décimal*—adding technical details of what was actually done, which neither Alder nor Gillispie covers adequately.

1.2 The birth of the metric system

Before the revolution of 1789, French units of weights and measures differed ‘not only in every province, but in every district and almost every town’ according to one contemporary commentator, Arthur Young; it has been estimated that ‘France contained a staggering 250 000 different units of weights and measures’ (Alder 2002

 © IOP Publishing Ltd 2017

pp 2–3). R D Connor explains that this was a legacy of the long duration of the feudal system in France, resulting in an 'utterly chaotic' system of weights and measures (Connor 1987 p 344). Clearly there was a strong case for standardisation. Furthermore, the official main French standards of weight and length were showing signs of wear and tear. They were, respectively, a set of weights known as the *Pile de Charlemagne* (owned by the French mint and used as a reference in the manufacture of coinage) and the Fathom of Peru *(Toise de Pérou)*, an iron bar, one-sixth of the length of which defined the *pied* or foot. Discrepancies were found when the 13 pieces of the Pile de Charlemagne were compared with one another, and the Fathom of Peru had replaced an earlier standard fathom which had been damaged, it in turn having replaced a still older standard for the same reason.

After an initial report by Charles-Maurice de Talleyrand-Périgord, the Archbishop of Paris, to the French Assembly, which proposed a system based on the length of a seconds pendulum (i.e. a pendulum with a period of 2 s), the Academie des Sciences set up a Commission of Weights and Measures to prepare the ground. They decided against a length standard based on the pendulum 'because of the arbitrary nature of the second' (Connor 1987 p 345) in any case, such a definition would have had to refer to a particular location, because of regional variations in the acceleration due to gravity, making it less universal. Instead they recommended a unit based on the earth's circumference.

In 1791, the Assembly defined two new units of mass and length, not in terms of arbitrary physical standards, but with reference to quantities which were regarded as 'taken from nature'. The *metre* was defined as one ten-millionth of the meridian arc from the North Pole to the equator, whilst the *grave* (later renamed the kilogram) was stipulated to be the mass of one cubic decimetre of distilled water at the melting point of ice. Hence the units of length and mass were to be based on natural measures—the circumference of the Earth and the density of water.

Basing the unit of length on the dimensions of the globe was not a new idea. An 'all-embracing decimal system' was first proposed between 1665 and 1670 by Gabriel Mouton in Lyon. The unit of length was to be based on the length of one minute of arc of a great circle, and the unit of mass linked to the unit of length and the density of water. However, this does not seem to have gone beyond a proposal; McGreevy notes that Mouton 'had not at his disposal the means for realising' his units (McGreevy 1995 p 140).

Royal approval for the metric system was obtained on 30 March 1791; it was decreed that 'it is necessary to decide on a measurement unit that is both natural and invariable'.

In practice, however, while the units were *defined* in such natural terms, it was clearly not going to be possible to refer back to the definitions every time a mass or length needed to be checked. A set of physical standards would, after all, be needed, the only difference being that these standards could, *in theory*, be checked against the definitions. What was needed, therefore, was a measurement of the meridian in terms of the old length standard (the *toise*), so that the old and new could be calibrated against each other, and new secondary standards made. Strictly speaking, of course, it was not necessary to *measure* the meridian: the meridian *defined* the metre, so what was really going on was a measurement of the *toise* ruler, in terms of the meridian, and hence the metre.

1.3 The Meridian Expedition

The measurement of distances of a few kilometres on the earth's surface is carried out by the method of *triangulation*, which consists of choosing a set of vantage points (or 'stations') on either side of the line to be measured, from each of which it is possible to view at least two others. Naturally, it was not intended to measure the entire quadrant from pole to equator. France extends from latitude 42.5° in the south to 51° in the north, or about a tenth of the quadrant; if this could be measured accurately, the length of the meridian could be calculated by scaling up, always allowing for the non-spherical shape of the earth. The Paris meridian covers this entire range, and could be extended by about another degree by measuring as far down as Barcelona in the south. Most of the meridian (from Dunquerque to Perpignan) had been measured before, in 1740 (Murdin 2009 p 34). Now it was planned to re-measure that section, and extend it to Barcelona.

Jean-Charles de Borda, described by Alder as 'France's leading experimental physicist' of the time, had recently produced his repeating circle, a device designed to measure angles very accurately. It consisted of two telescopes mounted on a common axis, so that they could be trained on two points simultaneously, and the angle between them read off. But it had an additional feature which enabled the user to double the angle repeatedly, creating an integer multiple, which, when divided by the number of multiples, reduced the uncertainty at the same time, making it a very precise instrument. It has been suggested that the existence of this instrument, together with the fact that its inventor occupied a very influential position in the hierarchy of French science, may have helped to sway the Commission away from the pendulum and in favour of triangulation (see e.g. Quinn 2012 p xxvii).

Two astronomers, Pierre-Françis-André Méchain and Jean-Baptiste-Joseph Delambre, were elected to carry out the survey. Each was to cover a section of the meridian, accompanied by three assistants and two of Borda's repeating circles, made by Etienne Lenoir. The previous measurement had been done by César-François Cassini (known as Cassini III—the third generation of a dynasty of astronomers) and indeed, his son, Jean-Dominique Cassini (Cassini IV) was the obvious choice to accompany Méchain; however, there were personal and political obstacles: Cassini's wife had recently died, and as a royalist sympathiser he was reluctant to participate in a project associated with what he considered an illegitimate régime. This was a wise decision, as it turned out; the mistrust and hostility shown to Méchain and Delambre would surely have been much worse in the case of a well-known royalist.

Because the previous survey had already identified suitable triangulation stations, the northern section was expected to be completed relatively rapidly. Hence the meridian was divided approximately in the ratio 2:1 from north to south, the meeting point being at Rodez, north-east of Toulouse. Delambre was to cover this section and Méchain the mountainous and uncharted southern portion. They set out in June 1792. The survey was supposed to take a year, but the astronomers did not return to Paris until 1798.

They were hampered by a number of factors. Accessing the stations often required them to be mountaineers and steeplejacks, and once the climbing was over they were

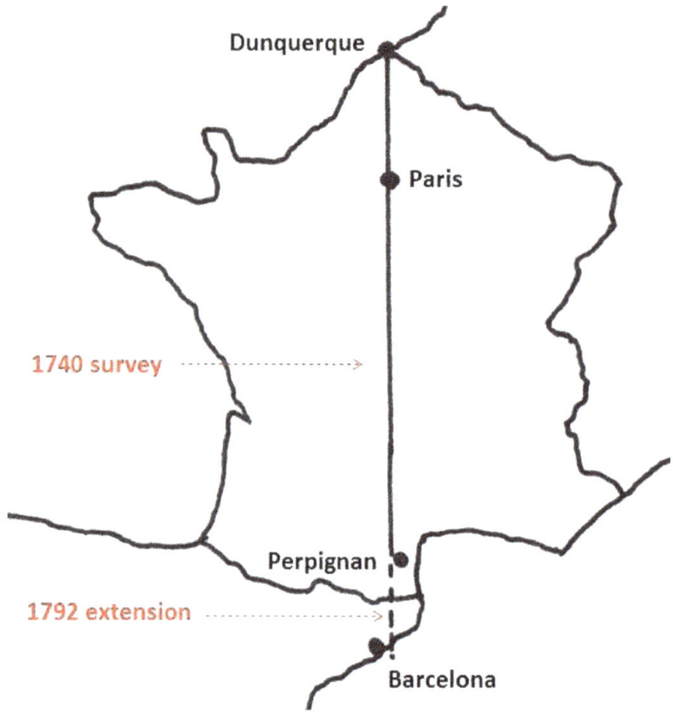

Figure 1.1. The 1740 and 1792 surveys.

at the mercy of the weather. Furthermore, Delambre found that many of Cassini's stations had deteriorated or been obscured by new developments. And they faced even greater dangers due to the volatile political climate of 1790s France: country people often mistook them for enemy spies, or worse, aristocrats. Méchain was detained by the Spanish in Barcelona during the war between France and Spain; and Delambre was arrested more than once, narrowly escaping the guillotine, and was at one point required to give an impromptu 'lecture' in the street, to explain the purpose of his expedition to an unruly crowd. They were not helped by the abolition of the Academy of Sciences, the body which had given them their remit, in August 1793; nor by the collapse of the currency, which rendered their financial resources almost worthless. Furthermore, there was an 18-month hiatus, during the period known as the Terror, when the expedition was called off altogether, before being reinstated in mid-1795. On top of all this, Méchain, who had suffered a serious accident while in Barcelona, became haunted by what he saw as an unforgiveable error in a latitude measurement, and became depressed and withdrawn.

1.4 How the Meridian was measured

Both Alder and Gillispie give comprehensive accounts of the expedition, but neither is very forthcoming on the technical details. For instance, both authors refer to 'spherical triangles', but neither explains what rôle spherical triangles might play in triangulation (see Alder 2002 p 24, Gillispie 2004 p 257). After all, the triangles that are actually observed are *plane* triangles, whose vertices are the signals mounted on the three stations, and whose sides are light rays; and light rays are straight lines, if one ignores refraction in the air (which is small in the case of light travelling close to the earth, and does not, in any case, produce a circular path), and bending by the earth's gravitational field (unknown at the time, and far too tiny to detect).

Triangulation

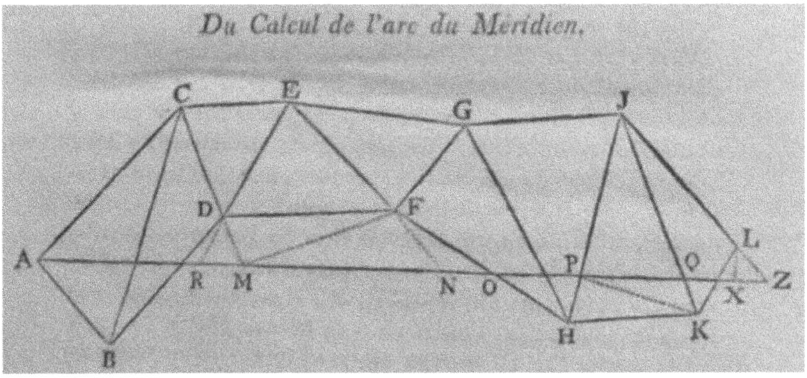

Figure 1.2. Legendre's diagram of a typical triangulation. The meridian is the line AZ. (Delambre and Legendre 1799). Courtesy of the Graves Library, UCL Library Services, Special Collections.

The basis of triangulation is to establish a grid of interconnected triangles, by means of a number of *stations* acting as vertices, covering the terrain to be surveyed. If an observer at any one station can see at least two others, the angle between the two can be measured, and if this is repeated over the whole grid, all the angles can be found. According to the laws of trigonometry, it is then only necessary to measure the length of *one* side of *one* of the triangles in order to calculate *all* the sides of *all* the triangles.

These triangles will not, however, lie in the same plane; on an irregular surface such as that of the earth, most stations will need to be elevated for visibility, and they will be at different heights. The triangles thus measured can be projected down onto a common plane surface. But the surface of the earth is not, of course, a plane. At a global scale on which local irregularities are unimportant, the earth's surface is represented by a body known as the *geoid*, a smooth oblate spheroid whose surface is at mean sea level (Bomford 1980 p 94). A *spheroid* is the surface generated by an ellipse rotated about its minor axis (Bomford 1980 p 643). In the case of measurement of the meridian, it is clearly a meridian of the geoid, and hence is curved; furthermore, the distance between lines of latitude will vary with latitude because of the non-spherical shape of the earth.

Isaac Todhunter, in his book *Spherical Trigonometry*, refers to three methods for finding the lengths of the sides of the grid. One is to use a theorem in spherical trigonometry, similar to the Sine Rule, which relates the sides of spherical triangles to their angles and the radius of the sphere. Another is to work with the *chordal* triangles, which are plane triangles joining the vertices of the spherical triangles; the third is to use Legendre's Theorem, which can be used to transform a spherical triangle into a plane triangle with sides equal to those of the spherical one. Todhunter says that Delambre used all three, and indeed, all are referred to in *Base du système métrique décimal*, although Delambre declares in the introduction to volume II that 'all my calculations were done using spherical trigonometry' (Delambre 1806 Tome II p xi, all translations from Delambre are the author's).

Before we launch ourselves into the minutiae of spherical trigonometry, it would be instructive to quantify the advantage of that system over plane trigonometry, by means of a specimen calculation. This is all the more important because the diagrams accompanying geodetic calculations invariably exaggerate the curvature of the triangles enormously, in order to make the geometry clear.

For an arc of a circle subtending an angle θ at the centre, if r is the radius of the circle, the arc length is $r\theta$, the chord length is $2r \sin \frac{1}{2}\theta$ and the length of the corresponding tangent (which touches the arc at its mid-point) is $2r \tan \frac{1}{2}\theta$.

Take, as typical, the portion of the meridian from Dunquerque to the line joining the stations at Watten and Cassel. This is given by Delambre as (to the nearest *toise*) 12 786, which is approximately 25 km. The radius of the earth is about 6378 km, so that, using the expressions given above, we find that the arc is about 16 mm longer than the chord and 32 mm shorter than the tangent. Consistently using chords will clearly lead to an underestimate of the length of the meridian; but not by much; perhaps 1 metre in the entire 283 km of the meridian, or about 4 parts per million. This probably would not have affected the final result, but nevertheless it is a testament to Méchain and Delambre's thoroughness that they chose the more precise, if more complicated, route.

The Borda repeating circle

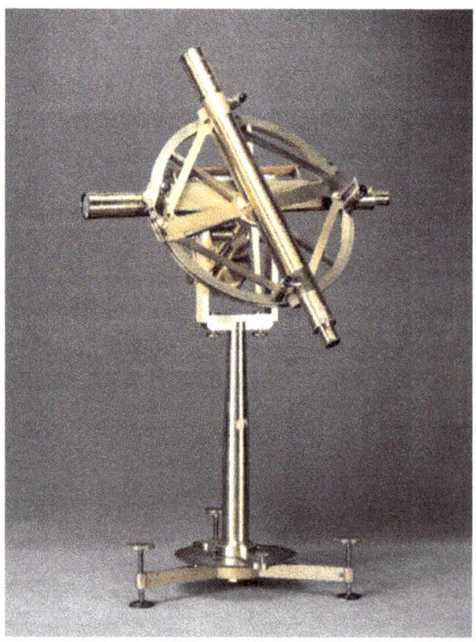

Figure 1.3. The Borda Repeating Circle. © Musée des Arts et Métiers, CNAM, Paris.

Borda's circle is a formidable instrument. It was based on an earlier invention by Johann Tobias Mayer (1723–62); Murdin says that Borda's contribution was to bring it to 'a sophisticated and practical state' (Murdin 2009 p 91).

Two telescopes were mounted, one above the other, on concentric brass rings, rotating against a fixed circular scale. The rings could be clamped together, or move independently. The geodesist would take it to one station and line up the two telescopes on the stations at the other two vertices of a given triangle, the top scope to the right and the lower one to the left. This generally necessitated tipping the circle as the two were unlikely to be in the same horizontal plane.

The angle between the two telescopes now represented the angle to be measured; then the two rings, clamped together, would be turned so that the lower scope was lined up with the right-hand station. The rings would be unclamped from each other and the upper scope turned back to the left-hand station, thus traversing an angle twice that to be measured. By repeating these actions, each time making a note of the multiple angle, the geodesist could reach an angle perhaps 20 or 30 times the desired angle. When this was divided by the multiple in order to obtain the original angle, the uncertainty would be reduced by the same factor. Thus an extremely precise reading could be obtained.

How precise? Well, Alder quotes Borda as claiming that 'an observer of sufficient patience should be able to eliminate nearly all error'; Cassini, when asked by the king to justify the new survey, stated only that it would reduce the precision of previous instruments from 15 s to 1 s (Alder 2002 p 56, 21). Borda's circle was actually calibrated in the new decimal degrees (100 per right angle) and, since it had a diameter of about a foot, the degree markings on the scale could have been no wider apart than 2.5 millimetres. One second, in the old sexagesimal units, would therefore be represented by a distance of less than a thousandth of a millimetre. Indeed, Méchain and Delambre's readings, as recorded in Delambre's report, are given to the nearest thousandth of a decimal degree, or about three seconds. Note that small telescopes were provided for reading the scale!

1.5 From angles to metres

Méchain and Delambre's handling of their data is interesting in itself. The meridian survey took place during a fascinating period in the development of data analysis. A few decades earlier, in 1750, Mayer (the original inventor of the repeating circle) had finally cracked the problem of processing large quantities of astronomical data—a problem whose resolution would seem obvious today but which had, only the previous year, defeated even Euler, acknowledged as one of the greatest mathematicians of all time (see e.g. Stigler 1986 p 27). Legendre, who worked in the metric commission alongside Méchain and Cassini, would announce his Least Squares method in 1805, closely followed by the ground-breaking achievements of Gauss and Laplace. But at the end of the 18th century it was still permissible to 'cherry-pick' data because they were considered more reliable, rather than simply ascribing them greater precision. We see this in the sample calculation given by Delambre for observations of the angle observed at Dunquerque between the next two stations,

Watten and Cassel. He records three series of measurements of this angle, taken at different times, and consisting of 20, 16 and 30 copies of the angle, respectively. After converting from decimal degrees to the older degrees, minutes and seconds, he averages the three overall results, carries out various adjustments, and then adjusts it again by half a second *en se rapprochant de la première série, qui est la meilleure*[1]. He does not justify this assessment of the relative merits of the three series. No uncertainties are given, even in the raw measurements.

After making allowances for refraction in the air, and the fact that, at any one station, the target and the observing point will not generally be the same place, or the target may be being viewed obliquely (Delambre gives detailed plans of many of the stations in his report) it is necessary to remove the random factor due to the inclinations of the observed triangles, which depend on the heights of the stations above sea level. This can be done by using basic trigonometry; the process is rather confusingly called 'reduction to the horizon', but actually consists of projecting the angles onto the *horizontal*, in other words a plane tangential to the geoid at a point vertically below the station[2]. This angle is clearly equal to the angle of the spherical triangle joining this point with similar points on the geoid, below the other stations.

Having performed this reduction, one is then in a position to apply the spherical version of the Sine Rule:

$$\frac{\sin A}{\sin(a/r)} = \frac{\sin B}{\sin(b/r)} = \frac{\sin C}{\sin(c/r)}$$

where A, B, C are the angles of the spherical triangle, a, b, c the lengths of the arcs making up its sides, and r the radius of the sphere. The formula (here taken from Todhunter (1871) article 41 p 19) is normally quoted for the case where $r = 1$, so that r does not appear explicitly; but I have included the radius in order to make it clear that a,b,c are distances and not angles.

This formula can be applied sequentially across the entire net, once at least one base line has been measured. In addition, at least one *azimuth* reading (the bearing of one station from another, with respect to north) is required, in order to position the meridian in the grid of triangles, and hence cut it into calculable portions.

1.6 Measuring the base lines

In theory, only one base line is needed; but Méchain and Delambre wanted to be able to cross-check their calculations. The beauty of having two base lines—ideally, one at either end of the net—is that each can be calculated from the other, and compared with direct

[1] 'Coming closer to' the first series, 'which is better'. Nowadays we might say he gives this series a greater *weight*. But no calculations are shown to support the adjustment (Delambre 1806, Tome I, *Mesure de la Méridienne*, p 11).

[2] This procedure is known as *reduction to the horizon* in both French and English (see e.g. Todhunter 1871 p 93). However, in French the word horizon has the additional meaning of 'landscape'. The horizon, in its everyday sense, is a line, not a plane, and varies with the observer's height above sea level.

measurement. So, two base lines were chosen: one at Melun, near Paris, and the other at Perpignan in the south. Both base lines were long sections of straight road (see figure 1.5).

Four special rulers, each two *toises* in length, were made for the baseline measurement; they would be placed end to end, and when all four had been laid, the first would be moved to the other end, and so on. Each ruler consisted of a strip of platinum, six *lignes* (or half an inch) wide and one *ligne* thick; fixed to the platinum strip at one end was a similar ruler made of copper, and about six *pouces* (inches) shorter. Because of the different coefficients of expansion of the two metals, the distance between the free ends of these rulers would vary with temperature, and could therefore be used to calculate it, and apply a temperature correction to the measurement, since the *toise* was defined at a temperature of 13° Réaumur. This end of the ruler also had a sliding *languette* (literally 'tongue') which had a vernier scale and could be pulled out to meet the adjoining ruler. Delambre does not actually explain why the *languette* was there at all; after all, one might imagine that the rulers could be lined up end to end without it. However, they were probably quite bulky and heavy items; the actual ruler was supported by a large block of wood, which rested on two iron tripods with spiked feet. Manoeuvring them into position must have been quite difficult, and attempting to put them in contact might have risked damaging them. Hence, one imagines, the *languette.*

Figure 1.4 shows a detail of the free end of the ruler. The end of the copper ruler can be seen, lying on top of the platinum ruler but not fixed to it at this end. The small copper strip is fixed to the platinum ruler, while the 'window' through which it is viewed is cut out of the copper ruler, and hence moves with it as the temperature varies. The vernier is used to read the position of the copper ruler relative to the fixed strip, and hence to the platinum ruler. The *languette* slides along between guides fixed to the platinum ruler. When the end of the *languette* is in contact with the end of the next ruler, the reading can be read from the vernier.

Further corrections were made to allow for any inclination of the base line; the Perpignan baseline, for instance, is represented in the report as an inverted V shape, consisting of two slightly inclined straight sections meeting at a summit.

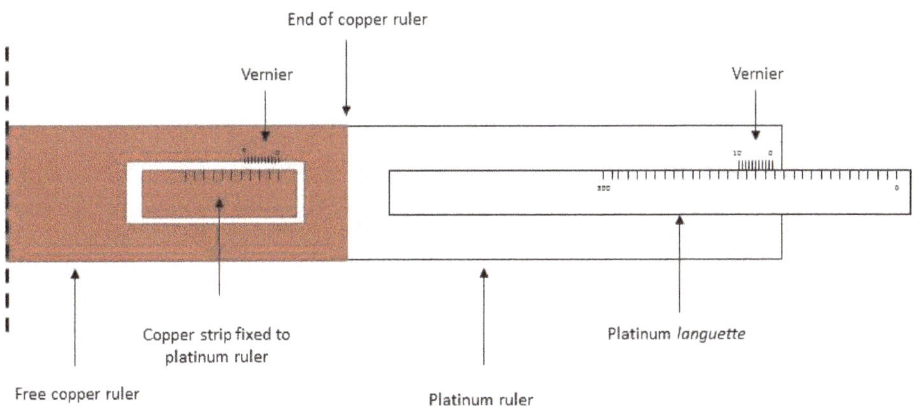

Figure 1.4. The free end of the ruler.

Figure 1.5. The southern baseline in 2017: the D900 road at Salses, looking towards Perpignan. Photo by the author.

1.7 Crunching the numbers

Delambre's results table for the Melun base measurement records the number of rulers used (that is, the number of 2-*toise* rulers that would be required to fill the base if placed end-to-end) as 3021, and the total of the readings on the *languettes* as 34.260 683 *toises*. These are added together, and, after various corrections for temperature, inclination, etc, the final base measurement is given as 6075.903 38 *toises*. He then discards the 0.003 38, explaining that this is done to allow for 'inevitable small errors in alignment, and the thickness of the rulers', so that his final figure is the 'round number' 6075.9 (Delambre 1806 tome II, p 45). For Perpignan, the final figure is 6006.247 848 *toises*. Again, there is a rounding, citing the same 'small errors', so that the final number is 6006.25 (Delambre 1806 tome II, p 55).

This process thus involved adding together some 3000 2-*toise* lengths and the same number of small *languette* corrections. The latter are given to 6 decimal places, i.e. to a millionth of a *toise*, or about 1/500 of a millimetre.

In order to provide a cross-check, Delambre calculated the length of the Perpignan base in terms of the measured length of the Melun base and the angles of the 53 triangles which linked the two together. He found the calculated length to be 6006.089 'demi-modules' (*toises*), and in comparing this with the measured length, given on this occasion as 6006.249 *toises*, found a difference of only 0.160 *toises*—'less than a thirty-seven-thousandth of the total, although the two bases are 33 000 demi-modules apart' (Delambre 1806 tome III, p 418).

Looking at the calculation with 21st century eyes, one might reserve judgement on whether such jubilation is well-founded. We would ask what the uncertainty was on that difference; if much less than 0.16 *toises*, we might not consider it a very good result at all, although of course it might still be sufficiently accurate for the purpose in hand. If the uncertainty was much greater, the agreement might be considered good, but the result less conclusive. But of course these modern methods were not available at the beginning of the 19th century; uncertainties were not quoted explicitly, but had to be gleaned from the precision with which the measurements were given; and the theory required to carry out uncertainty propagation was not yet available.

However, if one could be allowed to temporarily don a 'presentist hat', one might observe that the equation used to calculate one base (z) from the other (a)—a repeated application of the sine rule for spherical triangles—would look something like this:

$$\sin\left(\frac{z}{r}\right) = \sin\left(\frac{a}{r}\right)\left(\frac{\sin\theta_1}{\sin\theta_2}\right)\left(\frac{\sin\theta_3}{\sin\theta_4}\right)(\ldots)\left(\frac{\sin\theta_{2n-1}}{\sin\theta_{2n}}\right)$$

where the θ_i are angles of the intervening triangles, since, if n is the number of triangles, two angles of each are required. Application of the standard uncertainty propagation formula then yields, after approximating the sines and cosines of the angles to unity,

$$\left(\frac{\Delta z}{z}\right)^2 \approx \left(\frac{\Delta a}{a}\right)^2 + 2n(\Delta\theta)^2$$

where $\Delta\theta$ is taken to be representative of the uncertainty on each of the angles, and is measured in radians. Its value can be estimated from the fact that, as already noted, the angles were measured to the nearest thousandth of a *grade* (decimal degree), or about 2×10^{-5} radians. But what are the uncertainties in the bases? We can only assume that, since the base measurements each comprised some 3000 individual measurements, the value of 10^{-6} *toise* implied by the recorded data applies to each of these measurements, and these 3000 uncertainties will add in quadrature, giving an uncertainty of about ($\sqrt{3000} \times 10^{-6}$) or 0.000 055 *toises*. The fractional uncertainty on the base measurement is thus negligible compared to the contributions from the angles, so that, since $n \approx 50$, we arrive at a fractional uncertainty of about 2×10^{-4} for the calculated base, or (very approximately) 1.2 *toises*. We can conclude, on the basis of this very rough estimate (which has ignored all the other corrections etc, which can only increase the uncertainty) that quoting the calculated base length to three decimal places was perhaps a little optimistic. (In any case, in his description of the ruler, Delambre only claims to be able to discriminate to a hundred-thousandth (10^{-5}) of a *toise*, and to estimate halves, thirds and quarters of this amount; this is inconsistent with his results table, which suggests an ability to estimate tenths (Delambre 1806 tome II, p 3).

More significant would be the question of how uncertainties might have affected the meridian measurement (m) itself. Again, very roughly, we may say that this calculation runs along similar lines to the base calculation, with the angular uncertainties dominant, but, since at each stage a calculated length of a portion of the meridian is used to calculate the next one, the uncertainties mount up more rapidly: since the uncertainty in the *i*th portion will be of the order $\sqrt{2i}\,\Delta\theta$, the overall uncertainty is given by

$$\left(\frac{\Delta m}{m}\right)^2 \approx \sum_{i=1}^{n} 2i(\Delta\theta)^2$$

and hence

$$\frac{\Delta m}{m} \approx \Delta\theta\sqrt{2\sum_{i=1}^{n} i} \approx \sqrt{n(n+1)}\,\Delta\theta \approx n\Delta\theta.$$

Delambre gives a table listing the portions of the meridian which are cut off by successive triangles (Delambre 1806 tome III, p 47); this table is arranged in *geographical* order, from north to south, and hence starts with Dunquerque, whereas in reality the calculations must have started from the base line at Melun, some 25 triangles distant. That far along the chain, we ought to be looking at a fractional uncertainty of approximately 5×10^{-4}, or roughly 5 *toises*; yet all the intercepts are given in *toises* to three decimal places[3].

The final figure for the length of the meridian arc between Dunquerque and Mountjouy (a fortress near Barcelona, the southernmost point surveyed) is given as 551 583.6385 *toises* (Delambre 1806 tome III, p 77). This is an average of two values calculated in two slightly different ways; the discrepancy between these two values is 0.253 *toises*, which Delambre declares can be regarded as *insensible* (imperceptible). Indeed, since a modern analysis of the data would give a fractional uncertainty of approximately $50 \times 2 \times 10^{-5} - 10^{-3}$, or some 500 *toises*, the agreement is remarkable, but it is not clear whether anything can be deduced from the result about the precision of the method.

1.8 Extrapolation to the quadrant

At the extremities of the meridian, and at selected points in between, the astronomers also measured the latitude by finding the elevations of certain stars. However, because of the shape of the earth, extrapolating to the quadrant is not a simple matter of multiplying the length of the measured section of the meridian in *toises* by the ratio of a right angle to the difference in the latitudes of the extremities. The calculation was based on the assumption that the earth is a spheroid whose cross-section is therefore an ellipse, with its minor axis aligned with the axis of the earth.

The problem here was that the parameters of the ellipse were not known precisely. Delambre could, in the end, only derive a table of possible values of the metre in *lignes* for various values of the spheroid's flattening or *aplatissement*[4]. These values

[3] The use of parallel routes through the network of triangles allowed some of the distances to be measured by two or more independent methods, reducing the uncertainty; however, this reduction must be set against uncertainties in all the other corrections applied, which I have not considered in this example.

[4] The flattening f (or a as Delambre has it) is related to the eccentricity e of the ellipse, and to the major and minor axes a and b, by the formulae $f = (a - b)/a$ and $e^2 = 2f - f^2$. (Bomford 1980 p 646). Note that Alder confuses these terms; he *describes* the quantity referred to by Delambre as *aplatissement* and by Bomford as *flattening*, but *refers* to it as 'eccentricity'. The eccentricity of a body with $f = 1/300$ is about 1/12.

are given to five decimal places (implying an uncertainty of the order 10^{-5} *lignes*) but vary by as much as one sixth of a *ligne* as the flattening varies from 1/150 to zero. Delambre narrowed this range down to between 1/300 and 1/320, citing this as the most probable range in 'the general opinion of geometers and astronomers'; the table gave the metre as 443.304 59 and 443.309 85 *lignes*, respectively, for these values of the flattening, so he chose to propose to the Commission the 'round number' 443.3, truncating the decimals 'so as not to appear to affect a precision that we cannot justify' (Delambre 1806 tome III, p 103).

Nevertheless, the Metre Commission chose a different value. Méchain and Delambre had struggled to fit their data to a smooth spheroid, and different portions of the meridian, taken in isolation, predicted a different value for the flattening. In the end the Commission decided on 443.296 *lignes* as the length of the metre. This is very close to Delambre's figure for a flattening of 1/300. The choice was fortunate; nowadays the flattening is taken to be 1/298.25 (Bomford 1980 p 416).

This figure replaced the provisional value of 443.44 *lignes* per metre that Borda, Lagrange and Laplace had previously estimated on the basis of the 1740 survey; this had become the *provisional metre* on 1 August 1793 (Alder 2002 p 106). Now that the value was finalised, Lenoir was given the task of forging a platinum bar to represent the standard metre; it became known as the *metre of the archives*.

There are two striking things about this. Firstly, since the uncertainty in the flattening was surely known about at the commencement of the metre project, why go to all the trouble of making rulers that could measure to one millionth of a *toise* and a repeating circle capable of measuring to a thousandth of a degree, given that the final calculation could only be done to one part in 5000? And, come to that, why bother using spherical trigonometry when plane trigonometry could have delivered the required precision? What level of precision was the Commission seeking, after all? Even today, architects design buildings to the nearest millimetre, which suggests a precision of about one part in 10 000. Machine parts clearly need greater precision, but not in Méchain and Delambre's day: such parts were made to fit, not to specification. Could it be that the whole project was driven more by the pride of French geodesists and engineers than the need to define a new unit?

The second point is even more baffling. What the Commission did in the end was to *declare* a certain value for the metre in terms of the old measure; since both the flattening and the shape of the earth were so uncertain, they could attempt little more than an educated guess at its precise value. Doesn't the metre definition then amount, in the end, to more of an arbitrary stipulation—a value *guided by* nature, perhaps, but not *taken* from it? We shall see in the next section that the redefinition of the metre in the 19th century continued this trend, and abandoned the link to 'nature' altogether.

1.9 Conclusion

Ken Alder's subtitle is 'The Seven-Year Odyssey that Transformed the World'. But did it? Méchain and Delambre would have been appalled to learn that their metre 'taken from nature', which took seven years to calculate, would be gone in another

70; as we shall see in the next section, the redefinition by the Metre Convention reverted to the length of a physical standard, with no reference to the meridian. Yet, as we shall also discover, their work showed that using 'natural' measures was entirely possible, and kept the idea alive for long enough for the necessary technology to become available in the 20th century. Furthermore, the odyssey demonstrated the incredible precision made possible by the methods they used, particularly the Borda circle—methods which, as I have hopefully shown, were ahead of their time, awaiting the arrival of a more mature system of uncertainty analysis—and no doubt paved the way for the great European triangulations of the 19th century.

Part 2: The metre convention and the BIPM[5]

by Sophie Osiecki and Jim Grozier

1.10 The archive metre shows its age

During the time of the late 18th and early 19th centuries there was a demand for stable units of measurement for the purposes of map-making. The *metre des archives* was the primary standard against which others were compared. There was a method for checking—and this resulted in the metre bar being worn down at the ends. Outside states were rarely granted access to the original metre, and were instead directed to the secondary standard that belonged to the Conservatoire Nationale des Arts et Métiers (CNAM).

In the early 19th century there were several major triangulation exercises covering parts of Europe. Attempts to join together national triangulation networks often resulted in discrepancies. This was partly due to the existence of two kinds of length standard, which differed according to whether the length unit corresponded to the total length of the standard, or the distance between lines marked on it. The former was more susceptible to wear and damage, including that associated with the mere act of using the standard units, and were succeeded by the latter. But the archive metre was of the old end-contact type. A new standard was needed.

1.11 The international consensus: 1864–1875

In the 1864 International Conference of Geodesy in Berlin, attended by many European countries, the importance of the Metric system (i.e. a system of units for Europe) was stressed, and it was decided that there needed to be an international effort to create such a universal system of measurement. This had a marked effect on French pride, which had until then been at the centre of metrological matters, and the French government, Academie des Sciences and Bureau de Longitude (BL) all reacted. The BL checked what had been requested at the conference and stated that they could in fact do the necessary work (Quinn 2016). As such, they recommended the creation of an international commission.

The Great Exhibition in Paris, held in 1867, brought together scientific representatives of many nations. By this time, many countries had adopted the metric system, so that it was effectively becoming an international system rather than just a French system. Discussions took place about further standardisation, under the guidance of Academician Boris Semyonovich Yakobi, better known by his European name of Moritz Heinrich Jacobi, who was President of the Academy of Sciences in St Petersburg. He drew up a report recommending 'recognition of the metric system and the appointment of an International Commission of specialists to supervise the preparation of new standards'. However, the plans had to be

[5] This section is largely based on a talk given by Terry Quinn (Quinn 2016), augmented by Quinn (2012).

postponed because of the outbreak of the Franco–Prussian War in July 1870 (McDonald 1960 p 192).

The International Metre Commission was, nevertheless, established in Paris by the French Government in 1869, and met from 8–13 August 1870 at the CNAM. The two principal characters behind the establishment of the Metre Commission were Adolph Hirsch, the Director of the observatory at Neuchatel in Switzerland, and Wilhelm Foerster, Director of the Berlin observatory; these two were later to become key members of the BIPM.

Among the issues discussed by the Metre Commission was the question of what the new length standard should be based on: was it the existing definition in terms of the length of the meridian, or was it simply to be made to the same length as the Metre of the Archives? The second option then invited a further question: should it be the same length as the Metre of the Archives as it was then, or as it had been when it was made?

Due to the Franco–Prussian War and the insurrection of the Paris Commune and its bloody suppression, no further meetings of the Commission took place until 1872.

The meeting also created a Committee for Preparatory Research, which met from 2–14 April 1872, and discussed the variation of the thermal expansion of coefficients of standards and other technical questions, and further recommended the creation of an International Bureau of Weights and Measures. The Committee's recommendations were put to a full meeting of the International Metre Commission on 24 September 1872. This meeting decided on new international prototypes of the metre and kilogram, which were to be based on the metre and kilogram of the Archives in their present states. It also decided to create an International Bureau, based in Paris, and to ask the French Government to facilitate this. In November 1873, all member states of the International Metre Commission were invited to appoint representatives to a diplomatic conference in Paris. No date was set for this conference.

At the same time, preparations began at the Conservatoire des Arts et Métiers for the casting of 250 kilograms of platinum–iridium alloy for the production of standard metres and kilograms. This took place in 1873 and 1874, but was beset with problems: the alloy was found to be contaminated by small amounts of iron and ruthenium. The problems continued for several years; meanwhile the long-awaited diplomatic conference took place in Paris in the spring of 1875.

1.12 The metre convention: 1 March to 20 May 1875

Today the International system of units is still governed, more or less, as it was stipulated by the Metre Convention, a treaty signed in Paris in 1875. It was attended by so-called 'plenipotentaries' or in other words high-level diplomats from the participating countries, including representation from Europe and the USA.

From March to April, the scientists created two proposals: one recommended the permanent scientific institute as proposed by Hirsch, and the other recommended that a temporary institute be created.

There were, however, three countries that opposed the creation of the BIPM; these were Great Britain, Holland and Portugal. The British delegate was Sir George

Airy. He was in fact very much for the metric system but he was sent as the British delegate on the condition that he rejected the proposals made at the metre convention to create a permanent scientific institute. The British Government argued against the recommendation on three grounds. The first of these was that it was likely that the running costs would become too high, the second was that it would create an unfortunate precedent given that hitherto there had not been an international scientific institute. Finally, they stated that no-one in Britain was interested in the metric system[6].

The Dutch delegate argued from his own personal views that creating such an institute, which would require a high-level director, would mean that the individual in charge would have an overriding authority within science, which would have a negative effect on the development of science. He instead suggested that the institute be created temporarily, and on its completion it should simply be a place where the artefacts are kept, and the keys handed over to diplomats in Paris, which would be open to all those wishing to compare the standards to their own.

The third objection was from the Portuguese government, who argued that the creation of such an institute was unnecessary; there were institutes that already existed that would be capable of carrying out the functions of the one being proposed (such as the Bureau de Longitude and Conservatoire National des Arts et Métiers). It would therefore be much cheaper to use these rather than create a new one.

However, the majority agreed with what had been laid out by the preparatory commission, namely the creation of a permanent, international scientific institute: the BIPM.

Thus, on 15 April, the conference formally approved the establishment of a new international metrological institute, the Bureau Internationale des Poids et Mesures, and initialled the text of the Metre Convention, which was signed by the official delegates of the participating countries on 20 May.

The International Committee (the CIPM) had started work immediately after the 15 April meeting. The first priorities for the new body were to find a suitable site, appoint staff, and choose the instruments it would need to carry out its work. It had been offered the use of the Pavillon de Breteuil at Sèvres, a 17th century building that had been badly damaged by the Franco–Prussian war. The Pavillon was chosen over an alternative site at the Chateau de Compiègne, which at 75 km was felt to be too far from Paris.

Once these initial questions had been resolved, the International Committee considered the situation regarding the prototype metres and kilograms, which were still in the process of manufacture at the Conservatoire. The Committee was concerned about contamination of the alloy, and gave the French section, which was responsible for making the prototypes, two years to sort out the problems. This matter soured relations between the two bodies; the French section refused to supply samples to the International Committee, which then approached the French Government to resolve the impasse. Eventually the Conservatoire agreed to the

[6] Sir George Airy was not only in favour of the metric system, but also appreciated the need for an international agreement on units of measurement within Europe.

International Committee's request to order three new metres, and three new cylinders for kilograms, made by George Matthey in London, and not from the contaminated batch. Arthur Morin, the Director of the Conservatoire, had a foot in both these camps, as he was also on the International Committee and hence was caught in the middle of the controversy. He defended the contaminated standards until his death in1880.

The magnitudes of both artefacts—the International Prototype Metre and the International Prototype Kilogram—were based on their direct antecedents, the metre and kilogram des archives, respectively. Since then the definition of the metre has changed several times[7].

1.13 The metric system in the 20th and 21st centuries

Since 1875 there have been a total of 25 meetings at the General Conference on Weights and Measures. At the 26th meeting, scheduled for late 2018, it is hoped that the draft resolution, recommending that the system of weights and measures be based on defining constants of physics rather than material artefacts, will be adopted. The trend had been set by the redefinition of the metre at the eleventh General Conference of Weights and Measures on 14 October 1960 (Barrell 1962). The material artefact—the International Prototype Metre—was known to be unstable, and was redefined in terms of the fundamental physical constant, c, the speed of light in a vacuum. Now it is proposed that all the SI base units will be defined in terms of constants. Dr Terry Quinn, former director of the BIPM, remarked: 'I think that is the end, that is what the committee in 1791 were aiming at; [The new system of 2018] will be based upon constants of physics not related to any place or one nation.' (Quinn 2016). To drive home this point, Quinn translates the opening paragraph of the 1791 Report as follows:

> The idea of referring all measurements to a unit of length taken from nature was seized upon by mathematicians as soon as the existence of such units and the possibility of determining it became known. They saw it as the only way to exclude all that was arbitrary from the system of measurement and to conserve it unchanged so that no event or revolution in the world could cast uncertainty upon it. They felt that with such a system, belonging exclusively to no one nation, one could hope that it would be adopted by all.

Quinn adds: 'That was a great idea, and in fact it was the central foundation of the Metric System. But as you know they couldn't do it. But we are planning to do it in 2018.'

If there is anything to be taken from this brief survey of the history of the metre convention it is that things do not always go according to plan, and even when decisions are made, it can take rather a long time for them to be implemented. While

[7] Interested readers should consult www.bipm.org/en/measurement-units/history-si/evolution-metre.html for a brief overview of the main changes and Barrell (1962) for much fuller and fairly technical history of the development of the definition of the metre.

we remain optimistic about the present state of science and in particular metrology, there is always some room for caution. It seems that the resolution, whether or not it is formally adopted, will be a beginning of a new chapter in modern metrology.

References

Alder K 2002 *The Measure of All Things* (Abacus)

Barrell H 1962 The metre *Contemp. Phys.* **6** 415–34

Bomford G 1980 *Geodesy* (Oxford: Oxford University Press)

Connor R D 1987 *The Weights and Measures of England* (HMSO)

Delambre J B J and Legendre A M 1799 (?) *Méthodes Analytiques Pour La Détermination D'un Arc Du Méridien; Précédées D'un Mémoire Sur Le Même Sujet* (Crapelet)

Delambre J B J 1806 *Base Du Système Métrique Décimal Ou Mesure De L'arc Du Méridien Compris Entre Les Parallèles De Dunkerque et Barcelone* vols 1–3 (Baudouin)

Gillispie C 2004 *Science and Polity in France: The Revolutionary and Napoleonic Years* (Princeton NJ: Princeton University Press)

McDonald D 1960 *A History of Platinum* (London: Johnson Matthey & Co)

McGreevy T 1995 *The Basis of Measurement* vol 1 (Chippenham, UK: Picton)

Mirowski P 1992 Looking for those natural numbers: dimensionless constants and the idea of natural measurement *Sci. Context* **5** 165–88

Murdin P 2009 *Full Meridian of Glory* (Berlin: Springer)

Quinn T 2012 *From Artefacts to Atoms* (Oxford: Oxford University Press)

Quinn T 2016 *The Metre Convention and the BIPM* Lecture given on 17 March 2016, NPL, Teddington

Stigler S 1986 *The History of Statistics: The Measurement of Uncertainty Before 1900* (Cambridge, MA: Harvard University Press)

Todhunter I 1871 *Spherical Trigonometry* (London: Macmillan)

Chapter 2

From notion to precision: the SI second

by Rory McEvoy

In 1824 Royal approval was given for 'An act for ascertaining uniformity of weights and measures', within which a seconds pendulum was described as an instrument that should be used to attain a standard yard. Despite the importance of time in this definition of length, the SI second was not formally standardized until the mid-20th century. The 86 400 s that make up a mean solar day satisfied the demands of 18th and 19th century science. This chapter will outline a brief history of time determination and measurement, charting the key developments that rendered the second as a tangible and measurable unit through to the fast-paced technological developments in precision metrology of the 20th century.

2.1 Ancient times

Our basic concept of time stems from Earth's daily rotation and its annual revolution around the Sun. Prehistoric monuments such as Stonehenge in Wiltshire offer evidence that our ancestors may have placed the stones to serve as a calendric time-finding instrument in order to observe the passage of the seasons by the alignment of the stones with the Sun and Moon. These megalithic structures have been the stimulus for much study and conjecture; in relatively recent times an engineer proposed that there may have been a standardized unit of length, a megalithic yard, used by the builders of these monuments. However tempting they may seem, there is no way of proving any theory surrounding these pre-historic sites. As an extreme example, a team of researchers proposed that the key to the Wiltshire site was the bluestone and that it had been chosen for its resonant properties and the theory suggested that the monument was a long range communication device; in effect, it was a xylophone of Brobdingnagian proportions. These diverse theses, constructed on foundations of supporting evidence, may well be possible but will always remain conjecture. The anthropologist, Jacquetta Hawkes, caustically summed up her frustration with this area of research stating that 'every generation has the Stonehenge it deserves—or desires'.

doi:10.1088/978-0-7503-1487-9ch2

© IOP Publishing Ltd 2017

However, there are some better preserved prehistoric sites, such as the Chankillo temple complex in Peru, which contains a structure that bears very strong evidence that its purpose was to observe the time of year. The structure is formed of 13 stone towers, placed in a north–south line between two viewpoints from which the solstices can be observed as the Sun rises and sets alongside the extremities of the outermost towers. Because the Sun's passage along the horizon slows as it approaches the solstice, the towers could have been used to find the date to an accuracy of around a few days (Ghezzi and Ruggles 2007). The cultural importance of this monument is lost. There were, however, rare surviving aboriginal cultures and even rarer still, survival of their oral traditions, which held their society's relationship to time and calendar. Alexander M Stephen (c.1850–94) lived with the Hopi people in Arizona for three years before his death in 1894, documented, and in doing so helped to preserve, such a belief system. The Hopi's complex agricultural and spiritual calendar was regulated by observations of the rising and setting of both the Sun and moon against the mountainous Arizona skyline (McClusky 1977).

Such time-determination methods were potentially capable of dividing the year with some accuracy; however, the earliest tangible instrumentation for subdividing the day into hours originates from Egypt. The sundial was the principal method of dividing the day into hours and one of the earliest known dial plates with twelve divisions, dating to around 1300 BCE, was recently discovered in Upper Egypt by Swiss archaeologists in 2013. Another ancient Egyptian tool for finding the time was the Merkhet, essentially a plumb line that was suspended by an assistant. The observer, equipped with a notched stave, would align themselves so that the line appeared to intersect the pole star and the centre of the notch in order to observe the transit of bright stars behind the line. In this instrument are the basic requirements for optical time-derivation—the transit telescopes, found in time and longitude observatories from the 17th century onwards, follow the same principles: they are aligned true geographic north–south and are used to observe the moment of a star's transit across the observer's meridian. If the astronomer's purpose is to catalogue the relative positions of the stars, then a time-keeper is required to build the chart using two co-ordinates: the angular height of the star and the time of its transit across the observer's meridian. The Ancient Egyptians used water clocks (or clepsydra) to divide the night into hours.

The water clock saw continued use and some development in China. The well documented monumental water clocks by Chang Heng (born c 78 CE), I-Hsing (672–727) and Su Sung (1020–1101) used water-powered mechanical gearing to drive armillary spheres, providing a real time view of the heavens (Needham 1960). One of the armillary spheres on Su Sung's masterpiece was situated on an observation platform so that it could be used to observe the machine's error when the skies were clear and enable the keeper to correct the machine. It is of significant interest to the history of clock making in that it included a mechanical device that helped to smooth the rotation of the main driving water wheel by means of a mechanical stop–start device that was activated by the level of water in the buckets. This type of mechanism is known as an escapement and is the vital component and beating heart of any mechanical clock. Whilst there is some evidence in the last manuscript of Ismail ibn al-Razzaz al-Jazari (1136–1206) of transmission of knowledge from China to the West of this form of escapement, it does not bear any similarity to the early European clock escapements of the Middle Ages (Needham 1959).

2.2 The mechanical clock

The early history of the mechanical clock in Europe is somewhat obscured due to the broad meaning of the Latin word *horologium,* which could refer to any timekeeping device including sun dials, clepsydrae, sand glasses, candles, bells (both manually and automatically rung) and mechanical timekeepers. The first clear descriptions of mechanical escapements are found in two manuscript descriptions of clockwork planetaria by Richard of Wallingford (1292–1336) and Giovanni de Dondi (1318–1389). Both models were driven by a simple clock train with a mechanical escapement. Neither maker claimed invention of the escapement and, in both descriptions, it is treated as subordinate to the astronomical gearing with the suggestion that it was already an existing technology.

The descriptions of these two machines help to quantify the large disparity between astronomical knowledge and the precision of the mechanical clock. The *second* was very much a notional unit that astronomers could use in calculating astronomical cycles mathematically to great precision. For example Roger Bacon (1219/20–92) in his *Opus Tertium* (1267) defined the length of the lunar month in terms of hours, minutes, seconds, thirds and quarters. Richard of Wallingford's astronomical gear train was also very precise and calculated lunation with only a small error of 1.8 parts in a million, whereas the clock-drive to de Dondi's astraria required the user to frequently add or remove small weights to the oscillator, known as a foliot or balance wheel, in order to compensate for the inherent friction problems. Early examples of the foliot are, in their simplest form, a horizontal iron bar suspended at its centre enabling a rotary oscillation. The energy to maintain the oscillation is provided from the verge, which is fixed vertically to the foliot below the point of suspension and receives impulse from the train of wheels and regulates the motion of the clockwork. The verge and foliot is susceptible to changes in its driving force and any resistance in the train of wheels will cause the clock to slow its rate. De Dondi's machine incorporated 1800 engaging toothed wheels and so it is unsurprising that it needed constant attention to keep it in step with the heavens.

But how close were these clocks to being capable of measuring the second? Histories of clock accuracy often represent the story graphically with charts such as figure 2.1. This discussion will follow this graph's plotting of the history of accurate clock and evaluate the various milestones and where possible provide citation for the values represented. The first point on the graph (figure 2.1) concurs with many histories of timekeeping accuracy and suggests that the daily accuracy of these clocks was around 10^3 or around 16 minutes per day. This figure deserves to be treated with some caution as it coincides with the maximum seasonal difference between mean and apparent solar time—the 'Equation of Time'. Whether or not this similarity is a coincidence, this attributed precision can only be a broad

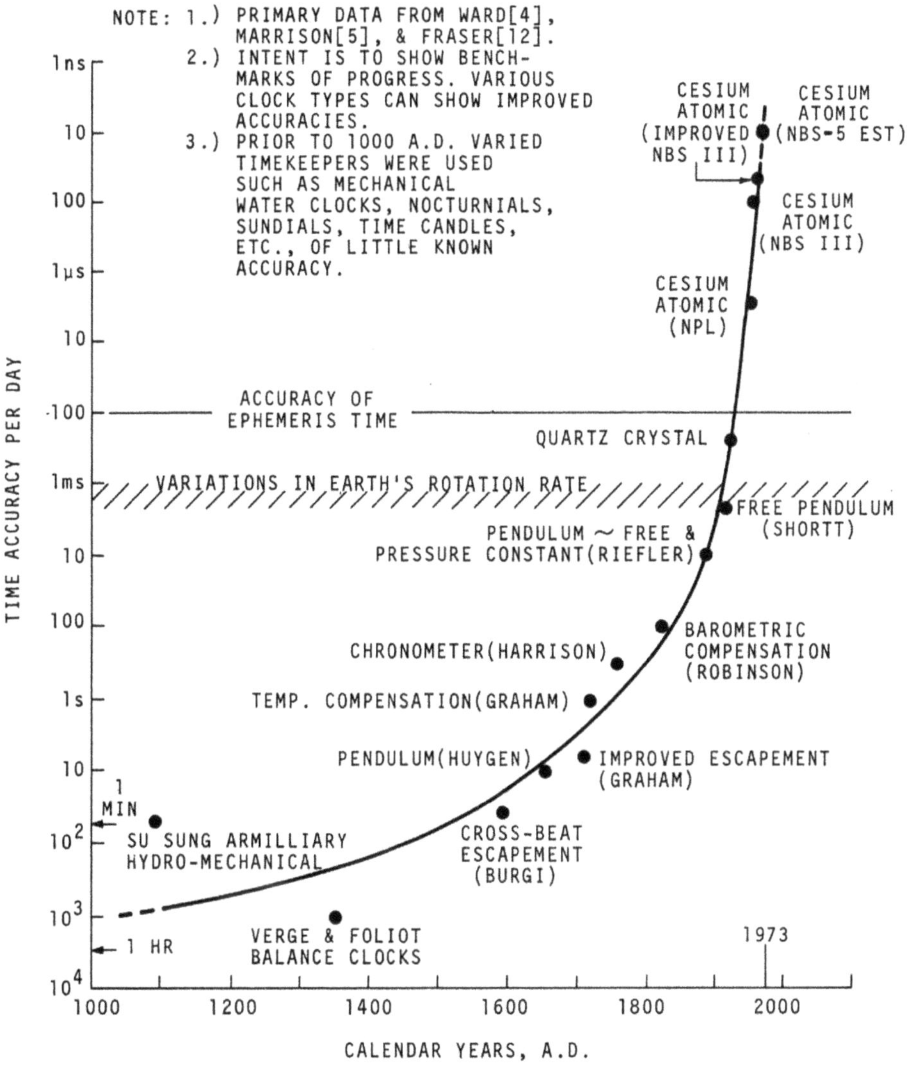

Figure 2.1. A typical chart of progress in timekeeping accuracy. Courstesy of The National Institute of Standards and Technology, from (Barnes 1974).

generalisation. A simple verge and foliot clock could have provided better accuracy but would have been reliant on constant attention to do so. One way to reduce the errors of such clocks was to use more than one and take a mean of the time indicated or identify any clock that was performing erratically. For this reason, astronomers used a mean of readings taken from several clocks when timing their observations.

The Danish astronomer Tycho Brahe (1546–1601) timed his observations by a number of clocks that indicated seconds and observed that as their driving weights descended they sped up. He described in a letter to Landgrave Wilhelm IV (1532–1592) that the acceleration in rate was caused by the additional mass of the rope, from which the driving weight was suspended, increasing as the weight descended.

Jost Burgi (1552–1632), employed as clockmaker to Wilhelm's court from 1604, overcame this problem by adding a device, known as a remontoire. In essence, the remontoire is a very short duration clock, requiring fewer engaging wheels, that is provided with an automatic re-winding mechanism. Wilhelm later reported that Burgi's clocks were able to keep time within one minute in a 24 hour period.

2.3 The pendulum

Galileo Galilei (1564–1642) was first to approach the precision required to measure the second in the laboratory. He observed that a pendulum swinging across a small arc appeared to be isochronous and went on to assert that its period was dependent on length and that this relationship was governed by a square law; that is, to double the period, the length needed to be quadrupled. In his experiments to determine the constant for natural acceleration, by rolling balls down an inclined plane, he used the pendulum to time the ball's descent. Simultaneously observing the swing of a pendulum and the position of a rolling ball over brief periods of time was simply not achievable so Galileo used the most precise instrument at his disposal—the weighing scale. By collecting water in a glass from a constant stream provided by a simple outflow clepsydra, Galileo was able to weigh the water collected over a set number of beats of the pendulum. He was able to judge relative accelerations by collecting water whilst the ball was in motion; the weight of water collected gave a precise quantity that represented time. However, this method only gave proportional changes for different distances and in order to quantify the period of his pendulum he needed the assistance of 'four patient and curious friends', who helped maintain and count the beats of the pendulum over a twenty-four hour period. Galileo and his friends counted 234 567 vibrations between consecutive transits of a bright star, which suggested that the period of the pendulum was around one third of a sidereal second (Drake 1978). Despite his efforts to quantify the period of the pendulum, Galileo used weight of water rather than units of time over distance to define natural acceleration in his *Dialogo* (1638). Subsequent and similar experiments were conducted by Marin Mersenne (1588–1648) and Gimabattista Riccioli (1598–1671) but they too found the quantification of the pendulum's period difficult. It was the Dutch mathematician, Christiaan Huygens (1629–95), who overcame the problem by successfully maintaining a pendulum by means of a mechanical clock.

Huygens' famous application of the pendulum to the mechanical clock had a pronounced effect on natural philosophy. His pamphlet *Horologium* (1658) described the pendulum clock and established priority of invention. From the pamphlet it is clear that Huygens saw greater reward in applying the pendulum clock to the 'science of Longitude … by taking to sea the most exquisitely constructed timepieces free from all error' (Edwardes 1970) Indeed, such clocks were produced for Huygens and his cohorts and taken to sea. After encouraging results, the opportunity for trial arose with an official voyage to Cayenne. Jean Richer (1630–96) was tasked with making observations of Mars from Guyana in an attempt to determine solar parallax and was given the subsidiary task of testing a number of experimental sea clocks. Huygens' aspirations were short-lived for the

Saint-Sebastien sailed into a storm not far from the French coast. The clocks were unable to withstand the violent motions of the ship and Richer elected not to continue with the trial.

However, when preparing to make his observations in French Guyana, Richer observed that his pendulum that had been rated in Paris, ran slow. To eliminate the possibility that this had been caused by accident, on re-calibrating the clock by astronomical observation, he made a simple seconds pendulum to be brought back for comparison in Paris on his return. His diligent actions were praised by Isaac Newton in the third edition of *Principia* (1726) as they highlighted the effect of reduced gravity on a pendulum clock and formed the foundation for study into the shape of the Earth using accurate pendulum clocks.

The pendulum clock saw further development in London along with the foundation of the Royal Observatory, Greenwich. Thanks to the patronage of Sir Jonas Moore (1627–79) the first English Astronomer Royal, John Flamsteed (1646–1719), was equipped with two extraordinary pendulum clocks made by Thomas Tompion (1639–1713) that incorporated the thinking and collaboration of some of the great minds of the era. The clocks followed a format that had been early demonstrated by Robert Hooke (1635–1703) in front of the Royal Society, where the swing of a heavy pendulum bob was maintained by a small pocket watch movement. As with Hooke's demonstration, Tompion's year-duration movements provided a very low-powered impulse to the pendulums, which meant that the movements were susceptible to failure through clogging with dust and debris. Recent experiments with replica movements in the Octagon room at the Royal Observatory showed considerable similarity to Flamsteed's experience with the originals. They ran very well after cleaning for about three months before the oil became clogged with dust. The enclosed space behind the wainscoting, depending on the conditions, acts like a chimney, drawing air and dust from the room through the clock movements or debris down from the whitewashed brickwork above. From Flamsteed's rating of the clocks against Sirius and the Sun from March to May 1677, we learn that the clocks ran with a precision of around 6 s per day.

When the clocks did keep 'good correspondence with the heavens' (Flamsteed 1677), Flamsteed was able to determine that the speed of the Earth's rotation was apparently constant before assembling his *Equation of Natural Days*—or as it became known later, the Equation of Time. This work, also carried out independently by Christiaan Huygens, was a milestone in time standardization and these tables were printed and distributed, often by clockmakers, as a means of setting clocks by the sundial.

Returning to the history of clock accuracy (figure 2.1) a small advance in precision timekeeping is credited to George Graham (c 1673–1751), successor to Thomas Tompion, for his improvement to the clock escapement. Graham did indeed improve the dead-beat escapement and set an enduring format for precision pendulum clocks. Though, it is arguable that the original motivation for producing the dead-beat escapement was ease of reading rather than greater accuracy as is commonly cited. This assumption may stem from the fact that it is almost always found in clocks used as a workshop, laboratory or observatory time standard.

Neither Tompion nor Graham claimed that this component gave greater accuracy or stability. Unlike the recoil escapement, which provides a constant back and forth motion to the seconds hand as it progresses around the dial, the dead-beat provides a positive quick motion, where the hand rests on the dial for the most part of a second before quickly advancing at the next beat. An overview of Graham's production of pendulum clocks shows that they feature both forms of escapement throughout the production period (c 1720–51) suggesting that the inclusion of the dead-beat was at the behest of the client.

The importance of legibility of the clock dial is particularly apparent when one considers the *Eye and Ear* method of observing star transits. This method was devised by James Bradley (1693–1762) and described by Nevil Maskelyne in *Greenwich Observations* (Maskelyne 1799). The method required the observer to glance at the transit clock, note the seconds before turning their eye to the telescope. Critically, the observer needed to continue to count the ticks of the clock and note the relative spacing of the transiting star (when the clock ticked) either side of a series of vertical wires seen through the eyepiece. This way a quick estimate of the time of transit past one wire could be made to a tenth of a second; then, taking an average from each wire, the timing of the transit could be averaged to give a value of a hundredth of a second.

The addition of temperature compensation, however, added significantly to the development of clock accuracy. Both George Graham and John Harrison (1693–1776) independently conceived solutions to the problem. Graham replaced the pendulum bob with a jar of mercury so that its upward expansion counteracted the lengthening of the pendulum rod in heat. Harrison's grid-iron pendulum used a ratio of the lengths of brass and steel rods, based on their relative coefficients of expansion, and arranged them in opposition so that they countered each other with temperature change. To get some quantification of the advantage brought by temperature compensation, a good record of the daily performance, or in the 18th century parlance, the going of *Graham 3*, the transit clock used at Greenwich by Nevil Maskelyne (1732–1811) to check the going of Harrison's *H4* shows that the clock regularly kept time to within a third of a second per day but occasionally experienced shifts of up to 0.8 of a second (Maskelyne 1767).

2.4 Pursuing precision

It is debatable as to whether or not Harrison's *H4* should appear on this graph (figure 2.1) at all. Being a portable timekeeper it belongs to an entirely separate branch of the horological taxonomy. It is a timekeeper, designed specifically for use at sea and therefore to cope with motion and extremes of temperature. All other components on the graph are designed for use on land and have the distinct advantage of remaining in a fixed position. Furthermore, *H4* is presented with a daily accuracy of better than one second per day, which will be drawn from the overall result of the second sea trial, when the timekeeper was in error by only +54 s after 156 days at sea (Gould 2013). The danger in this analysis of performance is that the hidden daily error could have been much less impressive than the mean. The subsequent published details of the watch's daily performance under trial at the

Royal Observatory were not so favourable, which according to Harrison was due to the fact that he had not had time to adjust it to optimum running condition. In its proper context, as a portable timekeeper, *H4* achieved a gargantuan leap forward in daily accuracy from the then ubiquitous verge pocket watch. The robustness and practicality of his design was eventually proven by the performance of Larcum Kendall's (1719–90) copy, known as *K1,* taken by Captain James Cook (1728–79) on his second voyage of discovery (1772–5). Following the return of *K1* after Cook's fatal third voyage it was entrusted to Thomas Earnshaw (1749–1829) for repair, during which time he supplied mathematician and Fellow of the Royal Society, George Atwood (1745–1807), with data on the construction of the balance and balance spring. Atwood added further credit to Harrison's work when he used the data in a paper, published in the *Philosophical Transactions* (Atwood 1794), remarking that 'It is always satisfactory to compare the motion of machines with the general laws of mechanics, whenever friction and other regularities are so far diminished as to allow reference to a theory'.

In terms of the application of maths to horological theory, a seminal paper on escapements in clocks and watches (Airy 1830) by George Biddell Airy (1801–92) set a standard by which clock and watch escapements were judged. Indeed, it was in this paper that the superiority of the dead-beat escapement was formally established, stating that it 'approaches very nearly to absolute perfection: and in this respect theory and practice are in exact agreement.' However, the paper concludes that the relatively new chronometer escapement might be better for clocks by virtue of the fact that it did not have the frictional engagement of the dead-beat, its manufacture did not require such skilled workmen and that the clock would never be out of beat. Airy collaborated with the clock making firm, Dent, to produce a pendulum clock with his design of chronometer escapement in 1869. The commission of *Dent 1906* was probably motivated by the need for improvement after reports of the extraordinary performance of an American clock by William Bond that had been in use at the Observatory in Bidston, near Liverpool since 1867. Both clocks represented a new era in precision timekeeping as both were stable and accurate enough to show definite and predictable responses to changing air pressure.

Reports on *Dent 1906* outlined the response of a 0.3 second loss per day for a one-inch increase in air pressure and by 1872 plans were already in place to fit a compensator (Airy 1872). The simple compensating device employed a standard j-tube barometer with a float. The barometer acted on the pendulum by means of a see-saw with a float on one end and fixed magnets on the other. The pendulum bob was also fitted with fixed magnets, so that an increase in air pressure would cause the mercury level to fall, bringing the magnets closer together and thus accelerating the clock's rate. The logical progression towards better timekeeping was to isolate the clock from all environmental conditions and house it within an evacuated chamber or one kept under constant pressure.

This was by no means a new idea and there were numerous precedents. Experiments and proposals for clocks running in a vacuum go back to the 17th century, however the technology was not available to both keep the clocks running and preserve the vacuum. One of the earliest and modestly successful attempts at this was a marine chronometer constructed by gunsmith Joseph Manton (1766–1835), which was trialled at the Royal Observatory from December 1808 through to the end of February, 1809 (Papers of the Board of Longitude 1784–1828). Manton's chronometer was wound through a stuffing box, and undoubtedly maintaining a constant pressure over long periods with such an arrangement was problematic. It was the German firm, Clemens Riefler, who were the first to make a successful form of tank regulator that could be isolated from changes in air pressure and re-wound electrically. The National Institute of Standards and Technology in the US reported that the limit of stability of their Reifler clock, purchased in 1904, was around 10 ms per day (Sullivan 2001). This new clock system was used as a time standard in many other time and longitude observatories worldwide; however, due to political sensitivities the Royal Observatory waited until 1921 to purchase one from Manchester University but, apparently, never put it to use at Greenwich. Instead the Observatory continued using *Dent 1906* as the sidereal standard through to 1922, when it was replaced by a copy of the Riefler system made by E T Cottingham (1869–1940).

Frank W Dyson (1868–1939) took on the role of Astronomer Royal at the Royal Observatory, Greenwich in 1910. Dyson's successor as Astronomer Royal for Scotland in Edinburgh was Ralph Sampson (1866–1939) who, unlike his predecessor, was actively interested in the study and improvement of precision clocks. Sampson was corresponding with railway engineer, William Hamilton Shortt (1881–1971), who was developing a new type of precision pendulum clock. In 1921 Shortt sent Sampson a record of his clock's performance, measured against the Paris radio time signal, which showed an unprecedented level of accuracy. When corrected for rate and a very small yet clearly defined relationship to temperature (0.008 s per day per degree centigrade change), Shortt's results showed an accuracy of within 0.01 s per day.

Unlike the earlier precision clocks, Shortt's system was free from the errors inherent in clock mechanisms and the characteristic behaviour of different escapements. The timekeeping element of the system was a pendulum with an Invar rod, housed in a near-evacuated sealed tank, which received an electro-mechanical impulse every 30 s. The electrical signal was provided by a servant pendulum clock which was periodically synchronized by return signals from the free pendulum.

Sampson wrote to Shortt in 1923: 'I shall not make a numerical report to you now, but merely say that the clock is unquestionably superior to Riefler and Riefler never turned out a better clock than the one we have here'. (Miles 2011) Sampson's numerical results and comparison of clocks at Edinburgh were communicated to the Royal Society of Arts in April, 1924 by Frank Hope-Jones (1867–1950) of the Synchronome electrical clock company, who were manufacturing Shortt's invention. Frank Dyson, who had chaired the meeting, wrote the following day to Hope-Jones requesting a quotation for supply and fitting of a free pendulum clock at the Royal Observatory, Greenwich.

The resultant history suggests that Dyson may have felt embarrassed by his successor's role in pioneering this new technology by virtue of the fact that Samson was not invited to lectures to the Royal Astronomical Society by staff from Greenwich on their experiences with the Shortt free pendulum clock. However, the published proceedings provide a good quantification of the performance of the free pendulum clocks. The performance of *Shortt 3,* in particular, was so good that the time indicated was referred to as mean sidereal time. Prior to the advent of these clocks the level of available accuracy in timekeeping was not sufficient to reflect the variations in astronomically derived time caused by the Earth's nutation. For the first time, the time kept by the sidereal standard was referred to as mean sidereal time and in order to correct to apparent sidereal time, the term coefficient of 0.08 s was applied over a six-month period (Jackson 1929).

Shortly before the publication of the going of the Shortt clocks at the Royal Observatory, the then director of the Strasbourg observatory, Andre Danjon (1890–1967), published his thoughts on how to redefine the standard second in order to increase the precision of astronomically derived time (Danjon 1929). The basis for Danjon's proposal was formed by various researches into the motion of the Solar system. He gave particular credit to the Dutch astronomer, Willem de Sitter (1872–1934), who had applied small corrections to the timing of historic observations of the moon and planets in order to resolve the results with their predicted motions. In essence, Danjon proposed to use the solar system as a clock dial, where the Earth acted as the main hand. Therefore, the redefined second would become a fraction of the tropical year rather than the Mean Solar day, and so moving towards an absolute and constant time scale that obviated the errors introduced by inconsistencies in the speed of Earth rotation.

2.5 Earth abandoned?

Since the 17th century Earth rotation, as measured by the astronomers through transit telescopes, had provided the ultimate time standard against which all other clocks were judged. Now astronomers had become acutely aware of the Earth's shortcomings as a timekeeper and, as was remarked in the 1930 paper on the Shortt clocks at Greenwich, the Shortt clocks were close to matching the Earth as a timekeeper.

The first palpable clue that the Earth may not have been the perfect timekeeper was presented by Edmund Halley (1656–1742) in his 1695 study of historic records of eclipses, in which he concluded that the lunar cycle was accelerating. The disparity between calculated timings of eclipses, based on historic records, and the actual observed positions of the moon were affirmed by English astronomer and surveyor, Richard Dunthorne (1711–75), in 1749 and the study was further developed by the French mathematician, Pierre-Simon Laplace (1749–1827), who conjectured that the Sun's influence over the moon was greater, causing the acceleration. It was not until the 19th century that this three body problem was fully understood (Thomson and Tait 1860). John Couch Adams (1819–92) improved on Laplace's mathematics and demonstrated that the Sun's influence on the moon was less by around a half of Laplace's figure, then French astronomer Charles Eugene Delaunay (1816–72) and American meteorologist, William Ferrel (1817–91) independently proposed that the inconsistency between the theoretical and observed positions of the moon was caused by the moon's gravitational pull and the resultant tidal friction that was slowing the Earth's speed of rotation (Jones 2000).

Naturally, this was not a complete solution to the problem. Better clocks and telescopes allowed astronomers to see that the moon still appeared ahead of or behind its predicted position, despite factoring in the deceleration of Earth rotation. In 1939 the tenth Astronomer Royal at Greenwich, Sir Harold Spencer Jones (1890–1960), having used several Shortt clocks over a three- year period to compare observations from the major time and longitude observatories, was able to categorically demonstrate the existence of an annual periodic variation in the speed of the Earth's rotation by demonstrating that errors in the moon's position were proportionate to those of the Sun and Mercury. Accordingly, extra to the deceleration of Earth rotation, there were seasonal fluctuations in the length of the day.

With this proof in place, for the astronomer, it was logical to downgrade meridional observations to a first approximation for time derivation, which would later be corrected by interpolation from a number of observations of the inferior planets (Mercury or Venus) and the Sun and Moon. But as Danjon acquiesced, it was still not a perfect system: the circular reasoning partly depended on theoretical celestial mechanics and, perhaps also in the light of Einstein's recent theory of relativity, the French astronomer had extra cause for such restraint. However, in the absence of better timekeepers, its adoption seemed inevitable and the Earth's days as our fundamental timekeeper were numbered.

2.6 Electronics appear

The Shortt clocks changed the paradigm in the observatory. Beforehand the Mean Solar and sidereal standards were checked and corrected by astronomical observations. The Shortt free pendulums, in conjunction with radio time signals, provided time by which the observations could be corrected. However, the supremacy of the Shortt free pendulum clock was fleeting and a new more accurate type of clock was already in development. The first quartz oscillators were produced in the late 1920s as frequency standards for use in the laboratories of the telecommunications industry and it was not long before they were adapted to measure time. The quartz crystal oscillator had a considerable advantage over the pendulum clock as it could be tuned to beat many thousands of times per second; any variation in the length of an oscillation would have a negligible effect on the timekeeping due to the smoothing achieved when averaged over one second.

The invention of radio signalling in the late 1800s played a significant role in catalysing fast-paced development in the time measurement, determination and, importantly, time standardisation. The first time signals were broadcast by the US Naval Observatory from New Jersey in 1904 and by 1910 both France and Germany were broadcasting time signals. This new conduit enabled observatories to compare their own results against the time signals and often revealed inconsistencies in the early days of the technology, which ultimately caused the foundation of the Bureau International de l'Heure (BIH) in 1913. The BIH continued to receive and correlate time signals from within the Paris Observatory. It remained unaffiliated to an official international body, due to the onset of WWI, but eventually became subsidiary to the newly formed International Astronomical Union (IAU) in 1920.

In 1942, the Post Office Radio Branch Laboratories at Dollis Hill, London provided a broadcast time that enabled the astronomers to use the signal to check a mean of quartz clocks. This mean would then be used to judge the rate of a single clock, which acted as the time standard. In broad terms, the Time Department of the Royal Observatory served two customers: navigators and astronomers, who required a time scale based on true Earth rotation, and frequency laboratories, who required a uniform timescale to work with. For both fields a separate monthly bulletin was published providing the corrections. By 1950, the increased precision of the optical instruments had revealed previously unobserved eccentricities in the Earth's motion. Humphry Smith (1914–2005) the then head of the Time Department, itemised ten corrections, excluding the most basic such as temperature effects on the optical instruments, that were applied to observed time. For this reason there was some criticism from engineers that the astronomical approach was too convoluted to be trustworthy and that the precision clock alone should be used (Smith 1950).

2.7 Independent standards

Louis Essen (1908–97) of the National Physical Laboratory, who would later become famous for developing the caesium beam atomic clock, expressed the opinion that the astronomically based time signals were not ideal for most scientific purposes and that it would be preferable to create a definitive standard that could conveniently be consulted in the laboratory (Essen 1952). His point of view is entirely understandable in the light of the fact that, at their 1950 conference in Paris, the IAU had agreed that time should be redefined as a fraction of the tropical year. The conference agreed a formula for converting the new unit to Mean Solar time, which would conform to the second on the sidereal date 1900.0 and that time reckoned by this unit would be known as *Ephemeris Time* (ET). Furthermore, the new unit was far from convenient. Spencer-Jones acknowledged Essen's standpoint in a comment responding to the articles (Jones 2000, Smith 1950) in the Institute of Electrical Engineers' journal by outlining that the formula required observations that might take up to two years to complete and enable the proper conversion.

Undeterred by the convoluted nature of deriving absolute time, the 1950 conference recommendation was put to the IAU in 1952 and the General Assembly adopted the resolution. By this time clock-making had advanced and the foundations for an atomic time standard were being prepared. In the late 1940s a team at the National Bureau of Standards in Washington DC were working on an advanced type of quartz clock that stabilized the frequency of the quartz crystal by means of microwave energy passed through a pressurized absorption chamber filled with ammonia gas. At 23 870 Hz the positon of the hydrogen atoms in the ammonia molecules inverted and in doing so absorbed the microwave energy. The reduced signal at the receiving end of the absorption chamber reported that the microwave energy was at its optimum level and the quartz crystal was adjusted to its correct frequency. The stability of the ammonia-controlled quartz clock did not surpass more than one or two parts in 10^8, which was the same as the best quartz clocks of the day and was still no better than the Earth as a timekeeper.

The next critical discussion on Ephemeris Time was at an IAU conference, convened in Dublin in 1955, just a few months into the successful and promising operation of the first caesium beam atomic clock at the National Physical Laboratory. Louis Essen attended the meeting, though not as an official delegate, and reiterated his feelings about the second of ephemeris time—that it was useless to engineers and physicists and that an atomic unit would be needed in future. His call for a postponement of the decision did not receive any support and the following year the International Committee for Weights and Measured (CIPM) formally defined the SI second as 1/31 556,925.9747 of the tropical year 1900. Essen remarked in his memoirs that *even scientific bodies can make ridiculous decisions* (NPL and Essen 2015).

For Essen the Dublin meeting was not entirely fruitless as he did garner support from William Markowitz (1907–98) of the US Naval Observatory, who proposed that the relationship between the atomic clock and ephemeris time be ascertained so that the atomic clock could then be used to broadcast astronomical time. The proposal was accepted, which left the door open for a future redefinition of the SI second as an atomic unit. Essen and Markowitz collaborated and using radio time signals they were able to compare the caesium beam atomic clock to ephemeris time as determined by Markowitz's moon camera to derive the current definition of the second—the time it takes for the 9 192 631 770 oscillations of the change in the energy state of a caesium133 atom.

2.8 Conclusions

The inevitable legacy of the definition of the SI second is somewhat complex. Different timescales, based on the SI second are made available, tailored to the needs of the recipient, be they astronomers, geodesists or navigators. The civil timescale, Coordinated Universal Time (UTC), is tied to Earth rotation and, of course, uses the SI second so there is a divergence between UTC and the atomic time scale (TAI). The International Earth Rotation and Reference Systems Service advises as and when the insertion of a leap second is needed in order to slow UTC and maintain the link with Earth rotation. This is in accordance with Humphry Smith's 1950 view that the *Earth, despite its vagaries must still serve as the fundamental natural standard of time.*

However, technology has moved on considerably since 1950 and precise time is essential to maintain the multitude of electronic connections worldwide. Given the volume of digital traffic, be it communications or commercial transaction, the leap second is a potential problem. Whatever time a leap second is scheduled it will always occur during the working day somewhere on the planet and there have been problems where computer servers have not been informed of an impending leap second and failed to function due to the illogical appearance of an extra second. Were such a failure to happen during the daytime and affect trading or safety systems the results could potentially be damaging. The question is currently being debated as to whether or not to abandon the connection to Earth rotation in our civil timescale and move to an atomic scale without leap seconds.

If this were to happen then the revolution would be complete, the clock would become master over astronomy. In just five millennia, the second has transformed from a notional quantity to the most precisely measured of all units and the clock that once kept time for the astronomer now points out the inconsistencies in the planetary motions.

References

Airy G B 1830 On the disturbances of pendulums and balances, and the theory of escapement *Trans. Cambridge Philos. Soc.* **3**

Airy G B 1872 *Observations made at the Royal Observatory in the year* **1872**

Atwood G 1794 Investigations, founded on the theory of motion for determining the times of bibration of watch balances *Philos. Trans. R. Soc.* **84** 119–68

Barnes J 1974 *Basic concepts of precise time and frequency* National Bureau of Standards monograph **140** 3

Danjon A 1929 Le Temps, sa définition pratique, sa mesure *L'Astronomie xliii* 13–22

Drake S 1978 *Galileo at Work his Scientific Biography* (Chicago: University of Chicago Press) p 399

Edwardes E 1970 Horologium, Christiaan Huygens, 1658. Latin text with translation *Antiquarian Horology* **7** 35

Essen L 1952 Frequency Standardization *Proc. IEE* Vol 99, Pt II, No 67

Flamsteed J 1677 *R. Soc.* MSS 243 (Fl.)

Ghezzi I and Ruggles C 2007 Chankillo: A 2300 year-old solar observatory in coastal Peru *Science* **315** 1239–43

Gould R T 2013 *The Marine Chronometer its History and Development* Antique Collector's Club **60**

Jackson J 1929 Shortt Clocks and the Earth's Rotation *Mon. Not. R. Astron. Soc.* **89** 239–50

Jones T 2000 *Splitting the Second* (Bristol: IOP Publishing) p 13

Maskelyne N 1799 Greenwich Observations

Maskelyne N 1767 An account of the going of Mr. John Harrison's watch, at the Royal Observatory, from May 6th, 1766, to March 4th, 1767

McClusky S C 1977 The Astronomy of the Hopi Indians *J. History Astron.* **8** 174

Miles R 2011 *Synchronome: masters of electrical timekeeping* Antiquarian Horological Society **177**

Needham J *et al* 1960 *Heavenly Clockwork: The Great Astronomical Clocks of Medieval China: A Missing Link in Horological History* (London: Cambridge University Press)

Needham J 1959 The missing link in horological history: A Chinese contribution *Proc. R. Soc. London, Series A, Math. Phys. Sci.* **250** 147–79

NPL and Essen R 2015 *The Memoirs of Louis Essen National Physical Laboratory* **68**

Papers of the Board of Longitude, Cambridge University Library Papers regarding the public trials and improvements of clocks and chronometers, 1784–1828 (RGO 14/24)

Smith H 1950 The determination of Time and Frequency *Proc. IEE* Vol 98, Pt II, No 62

Sullivan D B 2001 Time and Frequency a NIST: The First 100 Years 2001 *IEEE Intl. Frequency Control Symposium* **4**

Thomson W and Tait P G 1860 *Treatise on Natural Philosophy* section **403**

Chapter 3

Lord Rayleigh's determination of the ohm

Edward A Davis

Measurement of the unit of resistance by Rayleigh over 130 years ago was an extraordinary achievement, demanding considerable ingenuity and dedicated experimental effort. Rayleigh used two techniques in his determination of the ohm in absolute units—a rotating coil method devised by Kelvin and a spinning disc method conceived by Lorenz—with both yielding essentially the same result, and demonstrating that the values of existing British Association coils produced as standard resistors were over 1% too low. The 'mercury ohm', which at about the same time had become the accepted standard in laboratories around the world, was defined in terms of the length of a column of mercury having a resistance of one ohm. In a separate experiment, using his own resistance for calibration, Rayleigh determined this length to be 106.24 cm. An International Congress subsequently chose the rounded figure of 106.3 cm.

In this chapter, the above experiments are described in detail, followed by an outline of subsequent developments that has led to resistance being accurately represented via the Quantum Hall Effect. What will follow in 2019 is the redefining of Planck's constant and the electronic charge, which will result in the ohm becoming a base unit rather than a derived unit in the SI system.

3.1 Introduction

When, in 1879, the Third Lord Rayleigh took up the position of Professor of Experimental Physics in the Cavendish Laboratory, following James Clerk Maxwell in that post, he decided that his research would be directed towards determining electrical standards—a matter that at that time was becoming of increasing importance to the telegraph industry. Measurement of the unit of resistance became the initial focus of his studies.

Earlier in the century, various countries had suggested different standards for resistance: for example, in England one mile of No. 16 copper wire was proposed, in France 1 kilometre of 4 mm diameter iron wire, and in Germany 1 German mile of

doi:10.1088/978-0-7503-1487-9ch3 © IOP Publishing Ltd 2017

No. 8 iron wire. These proposals were largely unsatisfactory because the purity of the metals was neither specified nor controllable and their resistivities changed with temperature and stress. In 1860, Werner von Siemens, a German industrialist and founder of the company that still bears his name, proposed that the resistance of a column of mercury one metre long and one square millimetre in cross-section at 0 °C could be used as a reproducible standard. This became known as the Siemens unit.

In the electromagnetic system of units, resistance has the dimensions of length divided by time. This led to the idea of defining a unit of resistance in terms of a velocity, with Weber (whose name was later adopted as a unit of magnetic flux) suggesting one millimetre per second. However, this unit turned out to be too small for practical purposes. In 1861, the British Association for the Advancement of Science proposed the much higher value of ten million metres per second (10^9 in the cgs system of units), a value that became known as the BA unit. In order to relate this unit to the name of the discoverer of the law of resistance, the German physicist Georg Simon Ohm, it was also called *the ohm*. In 1791 the metre had been defined as one ten millionth of the distance along the Earth's surface from the pole to the equator measured along the meridian passing through Paris. So another way of expressing the ohm was one earth quadrant per second.

Figure 3.1 shows a standard resistor dating from c 1890 with its value expressed in units of velocity.

An experimental method used to determine the resistance of a coil in absolute units (i.e. ohms) was devised by William Thomson (later Lord Kelvin) and used by Maxwell, Stewart and Jenkin in 1863–4 at King's College London where Maxwell was then Professor. It involved measurement of the deflection of a small magnet suspended at the centre of a rotating coil, the resistance of which could then be determined from the angular deflection, the rate of rotation of the coil and its radius (see section 3.2). The work undertaken, under the auspices of a committee of the British Association, had led to the issuing of standard coils intended to have a

Figure 3.1. An early standard resistor with a value of very nearly 1 ohm expressed in terms of a velocity as seen on its label (courtesy of the University of Leeds Museum of History of Science, Technology and Medicine).

resistance of exactly 1 BA unit (ohm). However, in the following 16 years, measurements by others, using different methods, found departures from this standard by up to 2%—discrepancies that were unacceptable and required investigation.

3.2 The rotating coil method

Of the various techniques devised to determine the value of a resistance in absolute units (Lord Rayleigh 1882a), Rayleigh's initial choice was that used by Maxwell *et al* probably because the original apparatus and standard coils were in the Cavendish Laboratory when he took up his position there. However, he improved the method by which the rate of rotation of the coils was determined and treated much more carefully the matter of self-induction of the coil, as well as making corrections for temporal variations in the strength and direction of the Earth's magnetic field. He also designed and had built an improved version of the apparatus used previously.

Figure 3.2 shows the equipment used by Rayleigh for the experiments he conducted during the years 1881–1883.

Figure 3.2. Apparatus used at the Cavendish Laboratory for determination of the ohm in absolute units. Maxwell's original spinning coil sits on the table along with a synchronous motor. Rayleigh's improved apparatus, complete with stroboscopic disc, is on the floor, alongside equipment used to determine the ohm by the Lorenz method (see section 3.3).

Figure 3.3 is a photograph of the rotating coil apparatus as it now stands in the museum of the Cavendish Laboratory.

The closed coil, the resistance of which was to be determined, consists of many turns of copper wire. It was rotated at high speed about a vertical axis via a belt and pulleys using a paddle motor driven by water from a header tank on the roof of the building. A current was induced in the coil owing to its movement through the Earth's magnetic field. This current caused a deflection away from the meridian of a small magnet suspended at the centre of the coil. The deflection was measured via a light beam reflected onto a scale from a small mirror attached to the suspension wire close to the magnet. A simplified theory of the method is given in the box below, adapted from (Longair 2016).

Figure 3.3. Redesigned rotating coil apparatus used by Lord Rayleigh to determine the value of the ohm in absolute units. Its rate of rotation was controlled and measured with the aid of a stroboscopic disc attached above the pulley (see figure 3.4). Suspended in the vertical tube at the centre of the coil is a small magnet whose deflection from the magnetic meridian was measured by the deflection of a light beam reflected from a mirror attached to the suspension above the magnet. (Courtesy: the Cavendish Laboratory, University of Cambridge.)

SIMPLIFIED THEORY OF ROTATING COIL METHOD

The electromotive force, $\mathcal{E}$, induced in the rotating coil by the horizontal component of the Earth's magnetic field $B_{\parallel}$ is

$$\mathcal{E} = -\frac{\partial \emptyset}{\partial t} = \omega n \pi r^2 B_{\parallel}$$

where $\emptyset$ is the magnetic flux, ω the speed of rotation of the coil, n the number of turns in the coil, and r its radius.

The current induced in the coil is therefore

$$I = \frac{\omega n \pi r^2 B_{\parallel}}{R}$$

where R is the resistance of the coil.

The current I creates a magnetic field at the centre of the coil given by

$$B = \frac{\mu_0 I l}{4\pi r^2}$$

where $l = 2\pi r n$ is the total length of wire in the coil and μ_0 is the permeability of free space.

The effect of this field is to deflect a small magnet suspended at the centre of the coil through an angle θ given by

$$\theta = \frac{B}{B_{\parallel}} = \frac{\mu_0 I l}{4\pi r^2 B_{\parallel}} = \frac{\mu_0 \omega \pi l}{4\pi R}$$

In the cgs system $\mu_0/4\pi = 1$, so

$$R = \frac{2\pi^2 n^2 r \omega}{\theta}$$

Note that the Earth's magnetic field $B_{\parallel}$ has cancelled out.

It will be seen that the 'dimension' of the final formula for R is L/T, i.e. that of velocity.

The induced current is a maximum when the coil passes through the plane containing the magnetic meridian. It reverses its direction every half cycle. The magnetic field produced by this current also oscillates in magnitude but is always directed in a semicircle on the same side of the coil and at right angles to its plane. If the damping is sufficiently high, the suspended magnet adopts a steady angle of deflection.

The deflection of the magnet θ could be measured accurately enough with a sufficiently long light beam. The coil radius r was not that straightforward to determine because not only were there inner and outer radii owing to the finite dimensions of the windings but in addition the coil was split in the middle to allow for the magnet suspension and the axle. However, a suitable formula—involving several physical dimensions of the coil and its former—was derived to accommodate these complications. An ingenious method for determining the rate of rotation ω involved observation of a stroboscopic disc, shown in figures 3.2 and 4, through a pair of shutters attached to the ends of a tuning fork of known vibrational frequency. The tuning fork was electrically maintained (see figure 3.5). In order to obtain a

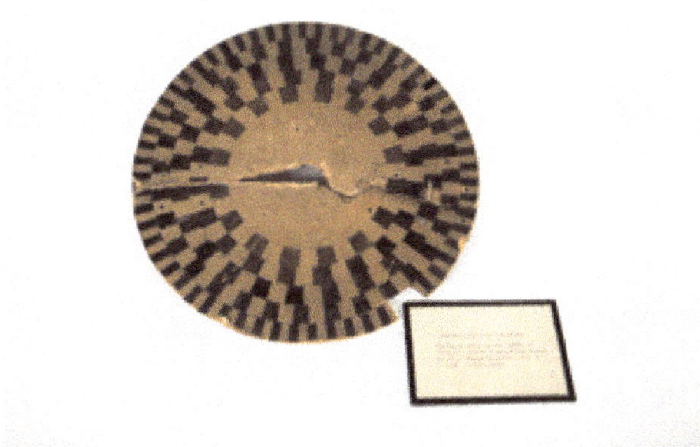

Figure 3.4. Stroboscopic disc used in conjunction with an electrically maintained tuning fork (see figure 3.5) to enable adjustment and measurement of rate of coil rotation. (Courtesy of the Cavendish Laboratory, University of Cambridge.)

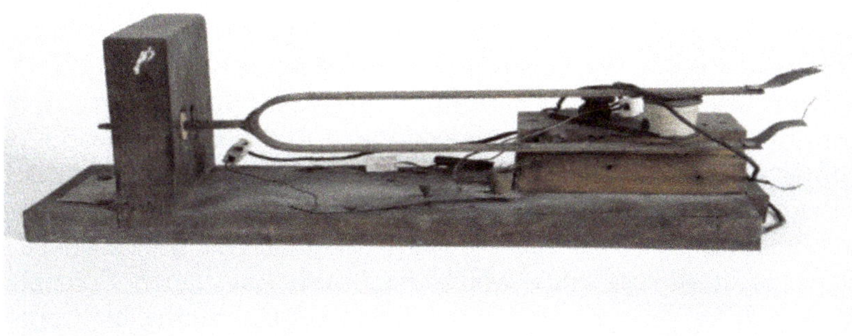

Figure 3.5. An electrically maintained tuning fork of the type used by Rayleigh to determine the rate of rotation of the coil. A dc voltage from a battery provides a current to the electromagnet (shown here with a white winding), which attracts the prongs of the fork. The circuit includes one of the prongs from which a wire is arranged to dip into mercury contained in a thimble (near the centre of the baseboard). As the fork vibrates, the circuit is opened and closed by this device, thereby providing an ac driving voltage to maintain the fork's vibrations. The stroboscopic disc is viewed through shutters (not shown here) attached to the ends of the fork. When a specific ring on the disc appears stationary, its frequency of rotation is some multiple of the fork's vibrational frequency.

perceived stationary pattern on the disc, final adjustment of the rotational speed was made by gripping the driving belt between finger and thumb and varying the applied pressure.

The simplified formula for R given above ignores several other factors for which Rayleigh had to correct in order to achieve the accuracy required. It is important to mention these so as to appreciate the care and effort needed in work of this kind. Self-induction of the coil enters into a full derivation, as does the magnetic moment of the suspended magnet and the torsion in the fibre suspension. Self-induction retards the phase of the current in the coil and reduces its maximum value. The correction for self-induction in Rayleigh's experiments was as high as 8%. It was calculated from the dimensions of the coil and also determined experimentally.

The Earth's magnetic field does not enter in the expression for R because, although a stronger field gives a larger induced current and hence larger deflecting field, this is counteracted by an increased restoring force on the magnet by the Earth's field. However, if the axis of rotation is not truly vertical, a correction is necessary for the angle of dip. Also, any changes in the magnetic declination during the experiment need to be compensated for; Rayleigh monitored this with an auxiliary magnetometer.

3.3 Value of the BA unit of resistance as determined by Rayleigh

After several years of painstaking work, during which the original apparatus used by Maxwell was replaced by an improved version, Rayleigh had determined the resistance of the coil in absolute units (Lord Rayleigh and Schuster 1881, Lord Rayleigh 1882b). This was then compared, using conventional bridge techniques, to standard British Association (BA) resistors, which were calibrated in multiples of the accepted international value of the ohm at that time.

Rayleigh's final result obtained in 1882 was that 1 BA unit = 0.986 51 ohms. The main reason for the BA unit being over 1% lower than the value obtained by Rayleigh was attributed to an incorrect value of the self-inductance of the coil in the original experiments of Maxwell *et al.* Furthermore, the principal error was traced to an unfortunate ambiguity in terminology used to denote the 'breadth' and the 'depth' of the coil. These physical dimensions had inadvertently been interchanged in the previous calculations.

Several other researchers around the world also found the BA unit to be smaller than the value of 10^9 cgs units, which the British Association Committee had intended it to have. In England, James Prescott Joule, during his work on the equivalence of mechanical work and electrical energy, deduced a value of 0.9873. In the light of mounting evidence, any coils sent into the British Association for calibration after 1883 were certified in BA units but also assigned alternative values in R ohms (real ohms), where 1 BA unit equalled 0.9867 R ohms—extremely close to Rayleigh's value.

3.4 The Lorenz method

Methods other than that of the rotating coil for determining the value of a resistance in absolute terms had been conceived and used. Indeed, Rayleigh himself made a comparison of several of these (Lord Rayleigh 1882a), concluding that the method proposed and used by the Danish physicist, Ludvig Lorenz, in 1873 was probably the best.

In the Lorenz method a circular metallic disc is rotated at a steady rate in a magnetic field produced by a current through a surrounding coaxial coil (see figure 3.6).

Two wires make contact with the disc at its centre and at its circumference. The voltage generated between the contacts is proportional to the rate of rotation of the

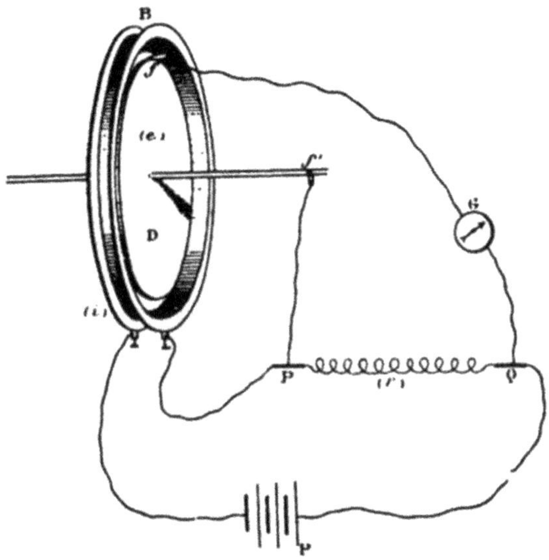

Figure 3.6. The Lorenz method for determining the ohm in absolute measure. The emf generated across the radius of the rotating disc D is balanced against that produced by the current through the resistance PQ, which carries the same current as that through the coils B. The value of the resistance at balance (null reading on the galvanometer G) is rM, where M is the coefficient of mutual inductance between the disc and the coil (calculable from the geometry) and r is the number of revolutions per second of the disc. Note: This figure is taken from Unites Electriques Absolues. Lecons Professees a la Sorbonne 1884–1885, translated by A Berget (Paris, 1889), p 154.

disc, the coefficient of mutual inductance between the coil and the circumference of the disc, and the current through the coil. If the coil current is also passed through a resistor—the value of which it is desired to determine in absolute measure—then, by choosing the rate and direction of the disc's rotation, it is possible to arrange for the voltage drop across this resistor to be equal and opposite to that generated across the disc. The null condition, registered on a galvanometer, occurs when $R = rM$, where r is the rate of rotation and M is the mutual inductance between the coil and disc. The former can be measured accurately and the latter calculated from the physical dimensions of the coil and the disc.

Rayleigh, in collaboration with Eleanor Sidgwick, his wife's sister, used this Lorenz method in 1883 to obtain a second determination of the ohm in absolute terms. The circuit diagram used—from the original paper (Lord Rayleigh and Sidgwick 1883)—is illustrated in figure 3.7.

The only portion of the set-up not mentioned above is the small loop circuit on the upper left. In this circuit, J–K represent two points on a stout copper wire, the separation of which could be varied to provide a voltage to compensate for two extraneous effects—firstly a thermoelectric voltage generated at the sliding contact on the disc and secondly a voltage produced by the vertical component of the Earth's magnetic field.

Rayleigh had at his disposal in the Cavendish laboratory two coils wound by a Professor Chrystal for R T Glazebrook who had previously used them for

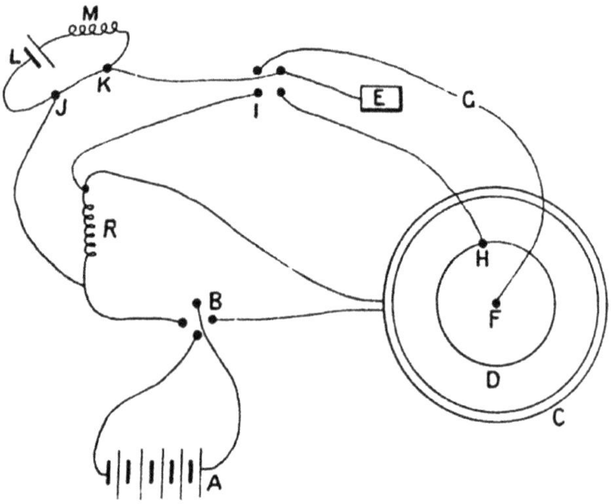

Figure 3.7. Circuit used by Rayleigh and Sidgwick in their determination of the ohm by the Lorenz method. Apart from the small circuit on the upper left (explained in the text below), the set-up is the same as in figure 3.6. B and I are reversing switches. G is a galvanometer, E is an Earth, FH is the rotating disc and C the coils.

measurement of the ohm by a different method. Rayleigh used these coils (see figure 3.8) in two configurations, one in which they were essentially in contact and another in which they were separated by a distance comparable to their radius. The rotating disc was made of brass and had a diameter roughly half that of the coils. Its axle was mounted vertically in the same frame that carried the coil in the apparatus described in section 3.2. This allowed employment of the same arrangements for driving the rotation and measuring the rate as were used in the rotating coil experiments. The coils producing the magnetic field were supported horizontally on wooden pieces screwed on the inner side of the three uprights of the frame shown in figure 3.3.

The mutual inductance M between the coils and the disc is to a first approximation equal, in cgs units, to $4\pi n(Aa)^{\frac{1}{2}}$ where n is the number of turns on the coils, A is the mean radius of the coils and a is the radius of the disc. However, modifications to this formula are necessary if the coils are separated. Also the inner contact is not the centre of the disc but the outer edge of the axle about which the disc rotates, reducing M accordingly. Suffice to say here that all corrections to the formula above involve length measurements and, to achieve the accuracy required, were measured with a precision of one hundredth of a millimetre. With regards to the rate of rotation, the stroboscopic disc method (see section 3.2) used by Rayleigh was ingenious and accurate but of course relied on an accurate knowledge of the frequency of the tuning fork used. The fork had a nominal frequency of 128 vibrations per second but calibration against an accurate clock pendulum revealed its rate to be 128.12 ± 0.02.

The final result from these investigations was 1 BA unit = 0.986 77 ohms, extremely close to the value he had obtained by the rotating coil method.

Figure 3.8. Pair of coils used by Rayleigh in the determination of the ohm by the Lorenz method. A brass disc is rotated in the magnetic field produced by the coils and the emf generated between its centre and circumference is balanced by a voltage drop produced across an external resistor. (Courtesy of the Cavendish Laboratory, University of Cambridge.)

3.5 The mercury standard

In the meantime, several laboratories had set up mercury standards of resistance. Werner Siemens had demonstrated 20 years earlier that mercury columns in glass capillary tubes could provide reproducible resistance standards with an accuracy of better than one part in 2000. In 1881 an International Congress of Electricians met in Paris for the purpose of establishing definitions of electrical units in a form suitable for enactment into legislation. The Congress proposed that the ohm be represented by a column of mercury 1 square millimetre in cross-section at a temperature of 0 degrees centigrade and that an international commission be charged with the task of determining by experiment the length of such a column. Clearly this could only be achieved by comparison of the resistance with some known standard.

Rayleigh undertook the task of determining this length using his own value of the ohm. In this investigation he was assisted again by Mrs Sidgwick (Lord Rayleigh and Sidgwick 1882). As with the rotating coil experiments—and indeed with any measurements requiring high precision—such determinations require considerable effort to eliminate possible sources of error. For example, any non-uniformity in the bore of the capillary tube had to be measured and taken into consideration. In addition, end corrections where the mercury was in contact with the electrical terminal cups were necessary. Rayleigh deduced that this latter problem was identical to that required in calculation of the pitch for the open ends of organ pipes—a solution to which he had given in his book *The Theory of Sound* some ten years earlier. The mercury itself was purified by distillation and treatment with nitric acid. A microscope fitted with cross-hairs controlled by screws graduated to one thousandth of an inch were used to measure the length of the mercury column. As the definition proposed in Paris specified 0 °C, all components had to be immersed in an ice-bath.

In 1882 Rayleigh published his finding that the length of a column of mercury one square millimetre in cross-section at 0 °C with a resistance of one ohm was 106.24(2) cm.

Two years later the International Conference for the Determination of the Electrical Units—the commission in Paris charged with setting the length of a mercury column as the resistance standard—adopted 106 cm. This was an average of values obtained by various people around the world, including that of Rayleigh. Furthermore, it was considered advisable for the value chosen to be correct to the last figure. The unit of resistance so defined was given the name 'legal ohm', although it was in fact never legalized. In Britain, William Thomson objected to the rounded value as he thought Rayleigh's measurements should be given more weight than those of others.

In 1891 in Edinburgh, a mercury standard of 106.3 cm was agreed upon by representatives from England, France, Germany and the United States and, two years later in Chicago, the above value was confirmed at an international congress under the presidency of Helmholtz. The name 'international ohm' was adopted and subsequently legal effect was given to it in the UK by an Order of Council. To quote the congress: 'The international ohm is *based* upon the ohm equal to 10^9 cgs system of electromagnetic units, and is *represented by* the resistance offered to an unvarying electric current by a column of mercury at the temperature of melting ice 14.4521 grams in mass, of a constant cross-sectional area and of the length of 106.3 cm'. The words in italics were clearly carefully chosen. Note that rather than specifying the cross section of the capillary tube, as was done previously, the congress specified the *mass* of mercury, thereby avoiding the difficulty of making a tube of exactly one square millimetre in cross-section.

Before concluding this historical chronology of events, it is of interest that in a paper from the National Bureau of Standards dated 1944, Curtis summarized findings from a previous decade of measurements (Curtis 1944). His conclusion was that an average of determinations from many countries resulted in the value: 1 international ohm = 1.000 494 absolute ohms.

This means in effect that 1 absolute ohm = 106.3/1.000 494 =106.2475 cm of Hg. It appears that it would have been better in 1891 if Rayleigh's value of 106.24(2) cm, as found by the rotating coil method, had been adopted as the standard for the international ohm, rather than 106.3 cm. It was correct to 1 part in 10 000.

3.6 Subsequent developments and modern resistance standards

The mercury ohm as the primary standard, to which all secondary standards were compared, remained in use until around 1920. Thermometer makers, as well as scientists, had contributed to its development and at the beginning of the twentieth century standards laboratories around the world all had their own mercury-ohm tubes. Implementation, however, was difficult, involving the filling of the tube with mercury for the resistance measurement and a refilling for the determination of its mass. Problems such as achieving uniformity of the bore of the tube, maintaining the temperature and allowing for end effects limited reproducibility of mercury standards to at best 20 ppm. On the other hand, a comparison of the values of wire-wound resistors constructed in the USA, England, France and Germany showed agreement with each other to within 10 ppm, with drifts of individual resistors often being less than 1 ppm. These secondary standards were normally made of manganin (84% Cu, 12% Mn and 4% Ni), the temperature coefficient of resistance (TCR) of which can be reduced to zero by proper heat treatment. Subsequently other materials were used, such as 'Evanohm', which also has an extremely low TCR and a very low thermal emf with the copper terminals. Modern examples of such resistance standards are shown in figure 3.9. The mercury ohm became redundant and was replaced in institutes holding standards by various other types of apparatus, such as

Figure 3.9. Resistance standards held at the National Physical Laboratory (NPL) England. Values available for calibration purposes range from 100 μΩ to 1 GΩ, with uncertainties of less than 1 ppm. Note the use of four terminals, two to carry a current and two to measure a voltage, thereby eliminating the resistance of the terminals. (Courtesy of the NPL.)

the Lorenz disc or carefully constructed inductance coils, which could provide primary determinations of the ohm in absolute terms (see (Curtis 1944)).

In 1960 the SI system of units was adopted through a resolution of the CGPM (Conference Generale des Poids et Mesures; known in English as the General Conference on Weights and Measures). In the SI system, seven *base* units are defined from which all other units can be *derived.* As far as electrical quantities are concerned, the unit of electrical current—the ampere—was defined, with the volt and ohm being designated derived units, as proclaimed in the following statement.

The ampere is that constant current which, if maintained in two straight parallel conductors of infinite length, of negligible circular cross-section, and placed 1 metre apart in vacuum, would produce between these conductors a force equal to 2×10^{-7} *MKS unit of force [newton] per metre of length.*

The volt is the potential difference between two points of a conducting wire carrying a constant current of 1 ampere, when the power dissipated between these points is equal to 1 watt.

The ohm is the electric resistance between two points of a conductor when a constant potential difference of 1 volt, applied to these points, produces in the conductor a current of 1 ampere, the conductor not being the seat of any electromotive force.

Practical *realization* of the ampere to high accuracy in accordance with this definition is obviously difficult, and current balances to do just that are only available in standards laboratories. The best current realizations of the ampere are in fact obtained through realizations of the watt, the volt and the ohm. A combination of any two of these three suffices. Realizations of the watt and the volt both involve balances in which electrostatic forces are measured in terms of mechanical forces. The ohm is realized using a Thompson–Lampard capacitor (see figure 3.10). These are all complicated techniques but the relative uncertainty in the value of the ampere obtained this way is currently a few parts in 10^7.

The resistance standard currently adopted is derived from the Quantum Hall Effect which gives a value of h/e^2 (h = Planck's constant; e = electron charge) to very high accuracy (von Klitzing 1986). It is extremely stable and reproducible. That this (so-named) von Klitzing constant corresponds to a resistance can be seen by the sequence of identities:

$$h/e^2 \equiv h\nu/e^2\nu \equiv \text{energy}/e^2\nu \equiv eV/e^2\nu \equiv V/e\nu \equiv V/I \equiv R$$

The von Klitzing constant, h/e^2, is not in itself an absolute measure of the ohm. However, it has been measured and compared to standard resistors in institutes around the world, providing a numerical relationship between the two—see for example the data in table 3.1.

In 1990 CODATA (Committee on Data for Science and Technology) decided on a value of $R_{K-90} = 25812.807$ ohms, together with a recommendation that this value be adopted internationally. Not surprisingly, this relationship yields, to a high degree of accuracy, the original definition of the absolute ohm, namely 10^9 in cgs units or $1\ \text{J s C}^{-2}$ in the S.I. system of units, when CODATA values for h and e are used.

The 1990 recommendation of the CIPM (International Committee for Weights and Measures) to *represent* the ohm by $(h/e^2) \times 25812.807$ did not constitute a

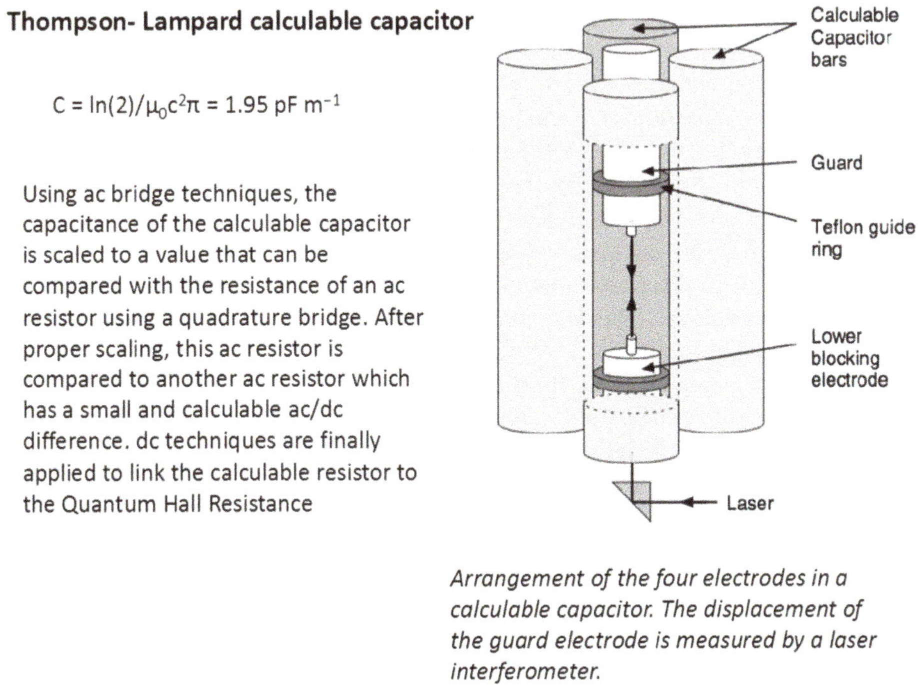

Figure 3.10. In 1956 the 'calculable capacitor' was conceived by Thompson and Lampard (Thompson and Lampard 1956). It offered a new method for determining the farad and the ohm in absolute terms. By the 1970s, secondary standards were being calibrated against calculable capacitors at the level of a few ppm.

Table 3.1. Values of quantised Hall resistance as determined by von Klitzing *et al* and at standards laboratories around the world up to 1988. The largest contribution to the error bars arises from uncertainties in the reference resistors used. I am grateful to Professor von Klitzing for his permission to use this table.

PRL <u>45</u>, 494 (1980)	25 812.680 ± 0.080 Ω
BIPM (Sèvres/Paris)	25 812.809 ± 0.003 Ω
PTB (Germany)	25 812.802 ± 0.003 Ω
ETL (Japan)	25 812.804 ± 0.008 Ω
VSL (NL)	25 812.802 ± 0.005 Ω
NRC (Canada)	25 812.814 ± 0.006 Ω
EAM (Switzerland)	25 812.809 ± 0.004 Ω
NBS (USA)	25 812.810 ± 0.002 Ω
NPL (GB)	25 812.811 ± 0.002 Ω

largest uncertainty due to uncertainty of the reference resistor!

redefinition of SI units. As the committee noted, to demote the ampere from a base unit to a derived unit in favour of the ohm would change the status of μ_0 as having an exactly defined value of $4\pi \times 10^7$.

However, the situation will change in 2019, when it is anticipated that the fundamental constants h and e will be fixed in value as part of the introduction of a new system of units following the abandonment of the existing artificial standard of mass, the kilogram. The watt balance—now called the Kibble balance after its founder Bryan Kibble—measures weight in electrical terms by balancing a gravitational force with an electromagnetic force. The latter involves measurement of a resistance and a voltage, which in turn can be calibrated by reference to the von Klitzing (h/e^2) and Josephson ($h/2e$) constants. The Kibble balance is at present being used to determine a value for h but once this is redefined and becomes a fixed quantity, the balance can be used in reverse to determine a mass, on the same footing within the SI system as length and time

Fixing the values of h and e will lead to a different value of $R_{K\text{-}90}$ than that which was adopted in 1990. The ohm, along with the volt, will then cease to be a derived SI unit. Its definition will be as originally conceived and its value fixed for all time in terms of two invariable constants of nature.

References

Curtis H L 1944 Review of recent absolute determinations of the Ohm and the Ampere *J. Res. National Bureau of Standards 33 Research Paper RP1* **606**

Longair M S 2016 *Maxwell's Enduring Legacy: a Scientific History of the Cavendish Laboratory* (Cambridge: Cambridge University Press)

Rayleigh Lord 1882a Comparison of methods for the determination of resistance in absolute measure *Philos. Mag.* **14** 329–46

Rayleigh Lord 1882b Experiments to determine the value of the British association unit of resistance in absolute measure *Philos. Trans.* **173** 661–97

Rayleigh Lord and Schuster A 1881 On the determination of the Ohm (BA unit) in absolute measure *Proc. R. Soc.* **32** 104–141

Rayleigh Lord and Sidgwick E M 1882 On the specific resistance of mercury *Philos. Trans.* **174** 173–85

Rayleigh Lord and Sidgwick E M 1883 Experiments by the method of Lorenz, for the further determination of the absolute value of the British association unit of resistance, with an Appendix on the determination of the pitch of a standard tuning-fork *Philos. Trans.* **174** 295–322

Thompson A M and Lampard D G 1956 A new theorem in electrostatics and its application to calculable standards of capacitance *Nature* **177** 888

von Klitzing K 1986 The quantized Hall effect *Rev. Mod. Phys.* **58** 519

Chapter 4

Temperature scales: past, present and future: 1700–2050

Graham Machin

4.1 Introduction

This chapter traces the development of reliable temperature measurement from the beginning of the 18th century to the present day. The potential scope of such a work is all but limitless so I restrict myself to describing temperature scales and the emergence of an understanding of what temperature represents, and I end with some suggestions as to what the future shape of thermometry might be after the Kelvin redefinition in 2018.

In the 18th century liquid-in-glass thermometers began to be produced in a number of countries in Europe, in most cases following a different proprietary scheme for indicating the temperature. By the beginning of 19th century, three scales; the degree Fahrenheit, degree Centigrade and degree Réaumur emerged as dominant. Temperatures on these scales, whilst they could be reliably measured, were arbitrary in the sense that the values had no underlying physical meaning. As such they are denoted *de facto* scales in this article.

Scientific understanding of the real meaning of temperature progressed, particularly in the middle of the 19th century through the development of thermodynamics by Rudolf Clausius, James Prescott Joule and William Thomson (Lord Kelvin). Besides that growing scientific understanding there was a parallel increasing practical need from industry to make more reliable measurements over wider temperature ranges. These two things led to the emergence in the early 20th century of *defined* scales whose temperatures were close to those given by thermodynamic methods but were practical and relatively easy to realise. Defined scales, disseminated to industry using simple, practical thermometers from National Measurement Institutes, have been the backbone of reliable thermometry across the world for around 90 years.

doi:10.1088/978-0-7503-1487-9ch4

© IOP Publishing Ltd 2017

Since 1954 the unit of temperature has been the kelvin, defined by assigning a fixed numerical value to the water triple point (273.16 K). In 2018 the units of the International System (SI) will be entirely defined in terms of defined values of a set of fundamental constants, and the kelvin will be defined by fixing the value of the Boltzmann constant. This fundamental change, coupled with the emergence of increasingly practical primary thermometry, may well lead to a slow *demise* of the current defined temperature scales and an increasing use of primary thermometry, for both the realisation and dissemination of the quantity temperature.

This chapter charts these three broad areas of development under the themes of *de facto* temperature scales, *defined* temperature scales, and the *demise* of defined temperature scales.

4.2 *de facto* temperature scales: 1700–1900

At the beginning of the 18th century a large number of temperature scales were in use in Europe. In this chapter they are designated *de facto* scales because the temperature values they gave were specific to fixed points chosen to anchor the scale and to particular material properties, namely the expansion of particular liquids in glass. As an example of the diversity of scales figure 4.1 shows a 1754 thermometer at the University Museum Utrecht, The Netherlands. This thermometer is remarkable in that it shows 18 different temperature scales all in use at the time.

Also illustrative of the diversity of temperature scales in use at the time, in George Martine's book of 1740 there is a chart comparing 15 temperature scales, which is reproduced by Chang (2004, p 161). Finally a 'conversion chart' reported by Goubert in the Bibliotheque National de France, reports no less than 28 different temperature scales in use around Europe at that time (cited by Gauvin (2012)) where a photograph of that remarkable chart can be found.

This confusing situation arose for a number of reasons, chief of which were the use of different fixed reference points (see box) for the scales, the fact that different numerical values were assigned to the same fixed points, and sometimes the scales were divided in different ways, leading to degrees of different magnitude. In addition, besides the fact that there was no physical understanding as to what temperature represented, the very nature of a fixed point was ill understood—leading, with hindsight, to some strange choices for fixed points, like the melting point of butter[1] or the temperature of the deep cellar of the Paris Observatory[2].

[1] Reported by Delancé in 1688.

[2] Reported by Hire in 1708. Hire was at the time responsible for Meteorological observations at the Paris Observatory.

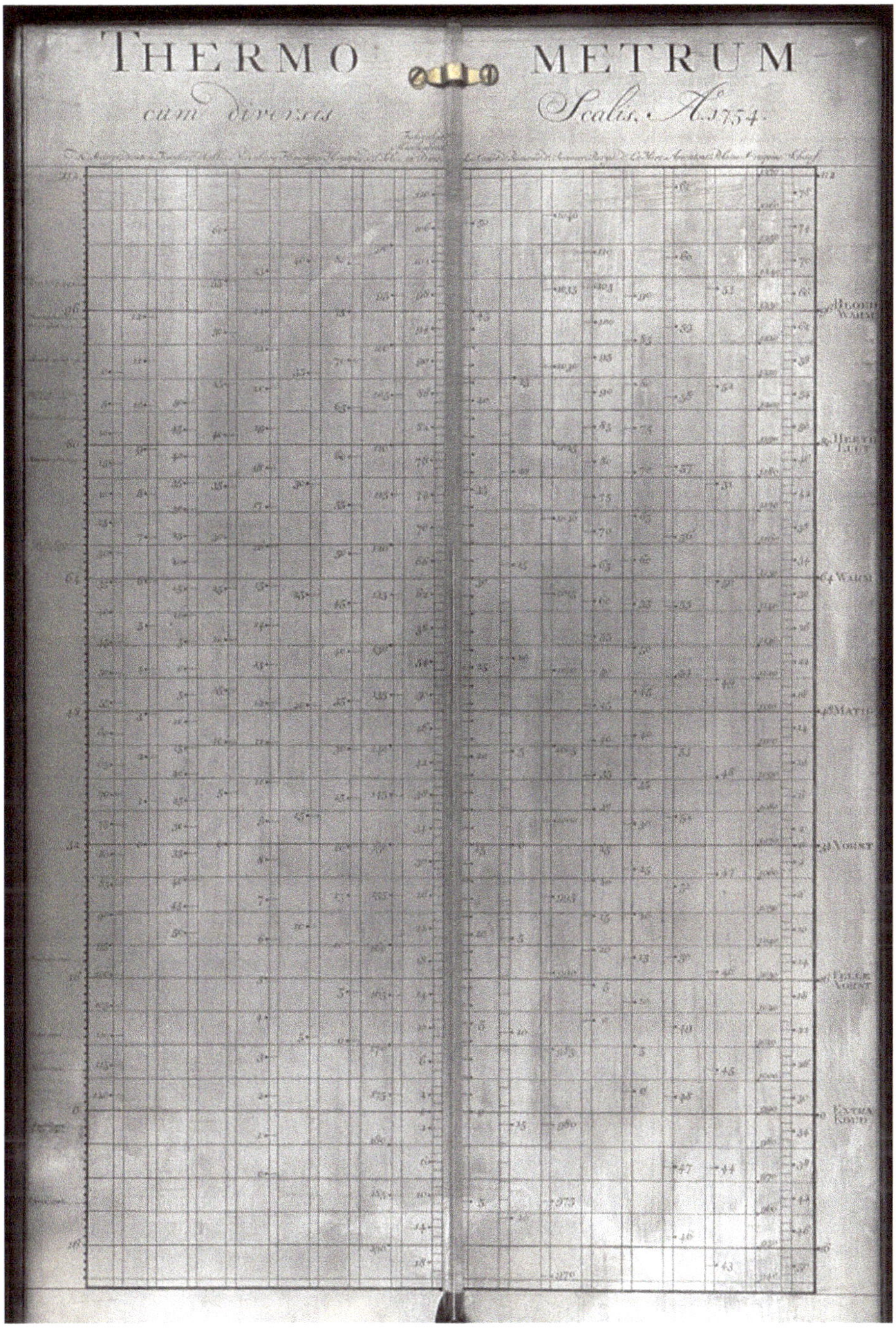

Figure 4.1. Thermometer from 1754 with 18 different scales indicated. Reproduced courtesy of University Museum Utrecht, The Netherlands (registration number: W11).

What is a temperature fixed point?
A temperature fixed point generally utilises the phase change of a pure material. Common phase changes that have been used for such purposes are the change from solid to liquid (e.g. melting ice), liquid to solid (e.g. freezing water), liquid to vapour (e.g. boiling water), or the triple point (the unique temperature where all three phases of matter, liquid, solid and vapour, co-exist). Examples of modern temperature fixed points, used in the current International Temperature Scale, are given in table 4.1. Fixed points are used by initiating the phase transition by any convenient means, for example by melting metal in a suitable container in a heated enclosure, and then inserting the thermometer (with suitable protection) into the fixed point material.

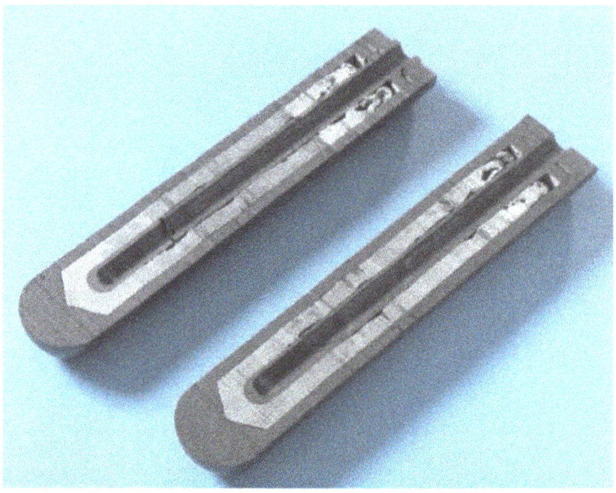

Figure 4.2. Section of a fixed point used for thermometry calibration.

Figure 4.2 is a section of a fixed point used for thermometry calibration. The shiny material is pure zinc metal which melts at around 419.5 °C (419.527 °C to be precise), the grey material is pure graphite. (Copyright NPL)

The outer graphite part of the fixed point, known as the crucible, holds the zinc as it melts and freezes. The inner part, known as the re-entrant well, is where the thermometer is inserted for calibration.

The phase transition is initiated by inserting the fixed point in a vertical furnace, heating the metal until it melts and then allowing it to freeze. As the material freezes it gives out heat (the latent heat of fusion) maintaining the entire ingot of metal at a constant temperature until it is all frozen. With care the metal can be held at its freezing point for 50 or more hours allowing for the calibration ofa number of thermometers during one freeze.

One example of an early scale from the UK is the Royal Society Scale from around 1720. Here the thermometer has an inverted scale in that 65° is the freezing point of water and 0° is the rather vague 'greatest heat' (approximately 32 °C), probably of the hottest weather. An example of a thermometer with this scale can be seen in the collections of the Museum of the History of Science, University of Oxford, a photograph of which is shown in figure 4.2.

By the end of the 18th century this confused and confusing situation began to resolve itself as three temperature scales came to dominate temperature measurement: the degree Fahrenheit, the degree Réaumur and the degree Centigrade, later Celsius. An outline of these scales is given below:

The Fahrenheit scale

This scale was described by Daniel Fahrenheit[3] (1686–1736) [°F] as reported in the Philosophical Transactions of the Royal Society (1724). He used mercury as the thermometric liquid, which has a lower expansivity than alcohol but a higher surface tension. This means that it does not wet the glass and so the thermometer has better repeatability. Fahrenheit's original scale had three defining fixed points; human body temperature defined the upper temperature at 96°, a mixture of ice and water made the intermediate temperature of 32° whilst a solution of brine made from equal parts of ice, water and salt (sea-salt or ammonium chloride) was the lower temperature of 0°[4]. This formulation did not persist, largely because the lower point was ill defined. Subsequently the °F was established using two fixed points; the water freezing and boiling points at 32 °F and 212 °F, respectively.

The Réaumur scale

René-Antoine Ferchault de Réaumur (1683–1757) [°Ré] described his scale in two long memoirs published at the beginning of 1730s. His devices used alcohol (i.e. the 'spirit of wine') as the thermometric liquid. However his scale as initially proposed was ambiguous but for different reasons to that of Fahrenheit. Whilst the lower fixed point of his scale was clear, it being the freezing point of water at 0 °Ré, his description of the upper fixed point was long winded and caused great confusion to his readers. He did not actually say that 80 °Ré, 'being a number convenient for dividing into parts', was the boiling point of water, though that was the interpretation of nearly all his readers. In the decades that followed the freezing point of water at 0 °Ré and the boiling point of water at 80 °Ré were adopted as the two fixed points used to establish the Réaumur scale. Later mercury, rather than the 'spirits of wine' was adopted as the thermometric fluid, and in this form the scale became widely adopted in Europe in the 18th century[5].

[3] Fahrenheit was born in Danzig but lived most of his life in the Netherlands

[4] Doubt has been cast on whether Fahrenheit really ever used this lower point as the use of sea salt or ammonium chloride would give two different temperatures. As Middleton says in his section on Fahrenheit's scale 'the natural tendency of an instrument maker [is] to wish to conceal his processes, or at least obfuscate his readers'.

[5] Réaumur's scale was sometimes referred to in 19th century literature for example as here in Tolstoy's War and Peace 'Christmas came and except for the ceremonial Mass, the solemn and wearisome Christmas congratulations from neighbours and servants, and the new dresses everyone put on, there were no special festivities, though the calm frost of minus twenty degrees Réaumur, the dazzling sunshine by day, and the starlight of the winter nights seemed to call for some special celebration of the season.'

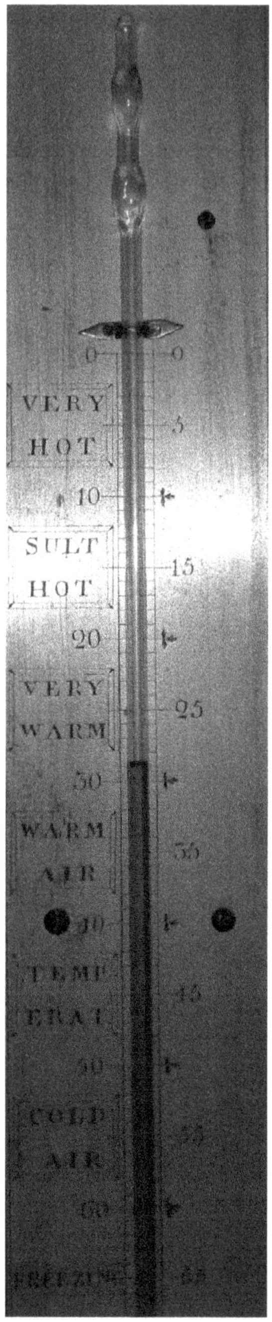

Figure 4.3. The Royal Society inverted scale, ice point at 65° and 'very hot' at 5°. Photograph by the author, reproduced courtesy of the Museum of the History of Science, University of Oxford inventory item 1 5341.

The centigrade scale

A slightly later scale was described in 1742 by Swedish astronomer, Anders Celsius (1701–1744) [°C] known as the centigrade scale. His proposed scale had two fixed points: the melting point of ice (snow) 0 °C and the boiling point of water at 100 °C. His thermometers used mercury as the thermometric fluid. One of Celsius' original thermometers survives at the Meteorological Institute of Uppsala University. The slightly idiosyncratic choice of temperatures for the fixed points was soon reversed by the famous Swedish botanist Carolus Linnaeus (1707–1778) and by others, possibly as early as 1745 and the scale was used like that thereafter. The scale became known as the Celsius scale, and this is now the official designation. Celsius temperatures and kelvin temperatures are related through the equation $t(°\mathrm{C}) = T(\mathrm{K}) - 273.15$.

For convenience and comparison purposes the three temperature scales, the Réaumur, the Fahrenheit and the Centigrade are all shown in figure 4.4.

More details on these and full discussions of earlier temperature scales are given in the excellent book by Middleton *A History of the Thermometer and its uses in Meteorology* (1966).

The 19th century situation and the emergence of absolute temperature

As the 19th Century progressed these three scales (measured with liquid in glass thermometers) progressively dominated practical thermometry, though little understanding as to the physical meaning of temperature existed. However, that was to change, mainly due to analysis of the efficiency of heat engines and the development of the laws of thermodynamics by Sadi Carnot, William Thomson (later Lord Kelvin) and others, and by insights provided by the application of statistics to atomic motion undertaken by leading figures such as James Clerk Maxwell, and Ludwig Boltzmann.

Thomson showed that there is a fundamental limit to the efficiency of ideal heat engines, which can be related to the temperatures of the hot heat source and the cold heat sink. By defining the temperature ratio to be equal to the ratios of the heats transferred, he was able to produce a scale which was independent of any material property (such as the expansivity of a liquid) and hence universally applicable. This work laid the foundation of a theoretical understanding of the inextricable link between heat (energy) and temperature.

The meaning of temperature

It seems impossible to believe now, but at the opening of the 20th century the existence of atoms was not yet firmly established. Nevertheless that did not stop scientists speculating on the nature of matter. For example Daniel Bernoulli in 1738 published his great work *Hydrodynamica*. In it he asserted that 'gases consist of a great number of particles moving in all directions … whose impact on surfaces causes gas pressure'. This inspired assertion is very close to our modern conception of gases but at the time was not accepted. It had to wait until the late 19th century

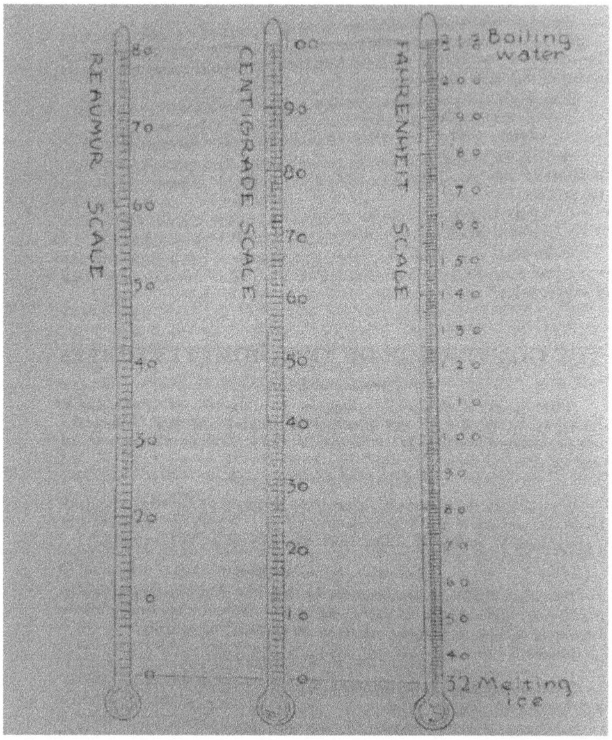

Figure 4.4. A comparison of the three main temperature scales in use in the 19th Century, Réaumur, Centigrade and Fahrenheit (diagram from Jameson 'The Thermometer and its Family Tree' (1914)).

when Maxwell between 1859 and 1866, building on the earlier work of Clausius, formulated a theory of the distribution of velocities of gas molecules, generalised later by Boltzmann. The formula, which describes the so-called Maxwell–Boltzmann distribution, allows for the calculation of the fraction of gas molecules moving at a specific velocity for any given absolute temperature.

Although beyond the scope of this chapter, essentially their work gave a detailed physical understanding of gases. That is, a gas consists of an ensemble of particles (typically atoms in inert gases, or molecules such as nitrogen, oxygen, carbon dioxide), which are travelling with a wide range of velocities and continually colliding and exchanging kinetic energy with each other. However, for a given absolute temperature the velocity of the particles follows a particular distribution (the Maxwell–Boltzmann distribution). The hotter the gas the more kinetic energy (and higher the velocity) the gas particles have *on average*, and the cooler the gas the less kinetic energy (and lower the velocity) the gas particles have *on average*. The kinetic energy (and velocity) become zero at a temperature of absolute zero (0 K) i.e. when the gas particles stop moving. This established a fundamental link between the atomically based statistical mechanics and the macroscopic thermodynamic approach of Kelvin and others. The link between these two approaches is that the thermal energy is seen to be proportional to the thermodynamic temperature, the constant of proportionality being Boltzmann's constant, *k*.

More details on the discovery of the meaning of temperature can be found in the book by Chang *Inventing Temperature: Measurement and Scientific Progress* (2004).

4.3 Towards *defined* temperature scales

Moving towards the end of the 19th century and entering the 20th century a clear understanding of the meaning of temperature had emerged. Researchers were engaged in performing temperature measurement according to those fundamental principles using devices now known as primary thermometers. These were almost universally gas thermometers and were used by, for example, determining how the pressure of a gas varies with absolute temperature in a constant volume (known as constant volume gas thermometry). However, such thermometers, and the temperatures derived from them, whilst being fundamental in that they reflected the underlying physics, were not in the least practical. They were slow, expensive, and often very large, usually requiring a laboratory full of equipment to make measurements.

In contrast, the art of making practical thermometers had so advanced that temperatures could be measured very reliably using any of the *de facto* scales. In addition, new types of practical thermometers were being developed, suited for use in the emerging industrial age, over wide temperature ranges. One which came to dominate the measurements of temperature in industry was the thermocouple[6]. Another, also of great importance, was a thermometer based on the electrical resistance of a platinum wire, first proposed by William Siemens in 1871. But it was left to Hugh Longbourne Callendar and colleagues to refine and promote the use of such devices. While at the Cavendish Laboratory, University of Cambridge, Callendar wrote a thesis on the subject which led to him being elected a Fellow of Trinity College in 1886. His development and promotion of platinum thermometers led to proposals for a practical scale of temperature based on a platinum resistance thermometer, though it was some years before it became a reality. More details of Callendar's work can be found in the book *A History of the Cambridge Scientific Instrument Company by Cattermole and Wolfe* (1987).

There was growing pressure not only from research, but also from the practical needs of industry to make temperature measurement that was both reliable and also linked to absolute (thermodynamic) temperatures.

This need was ultimately met through the development of *international temperature scales*, that is temperature scales whose temperatures were close to thermodynamic (absolute) values and yet simple to realise and, importantly from an industry and trade point of view, established on a worldwide scale. At the heart of such scales were fixed points of pure materials [see box]. The thermodynamic temperature of

[6] A thermocouple is a deceptively simple thermometer made of two dissimilar wires joined at the measurement junction. The sensor develops an electromotive force that is related to the temperature (gradient) it experiences. There are a wide variety of thermocouples in use, the most reliable being those based on platinum and alloys of platinum with rhodium. For example the so-called type R thermocouple is made from a wire of pure platinum and another wire of platinum alloyed with 13% rhodium.

such fixed points, be they the melting/freezing point of metals, or the triple points of water or gases, were determined using primary thermometers so that their 'true' thermodynamic temperatures were known with low uncertainties.

These temperature fixed points could then be used to calibrate relatively simple thermometers, such as those developed by Callendar, establishing a robust relationship between the measured quantity (e.g. electrical resistance) and the thermodynamic temperature of the fixed points.

The first truly international temperature scale was established through the auspices of the Bureau International des Poids et Mesures (BIPM)[7]. Building on earlier work by Pierre Chappuis at BIPM, and that of Callendar and others, led to the adoption of the International Temperature Scale of 1927, the so called ITS-27. The ITS-27 consisted of three temperature ranges:

- −190 °C to 650 °C: a platinum resistance thermometer was used as the interpolating instrument and calibrated at the boiling point of oxygen (−183 °C), the freezing point of water, the boiling point of water, and the boiling point of sulphur (444.6 °C).
- 630 °C to 1063 °C: a thermocouple based on Pt and Pt 10%Rh alloy (known as type S) was used as the interpolating instrument calibrated at the freezing points of antimony (630 °C), silver (960 °C) and gold (1063 °C)
- above 1063 °C: an optical pyrometer (measurement of 'thermal radiation' emitted by a black-body) was used

The scale was thus established on these three devices, the platinum resistance thermometer, the type S thermocouple and the optical pyrometer. These were then used to calibrate other sensors or artefacts, which in turn were then used to transfer the ITS-27 to industry, science and any other user who needed to determine reliable traceable temperatures.

This pattern has then been followed during the 20th century and into the 21st to establish a worldwide measurement infrastructure for reliable temperature measurement. Defined scales of increasing sophistication were constructed and adopted, each one realising thermodynamic temperature more closely than the one it succeeded. In brief, the scales (all of which had, each time, refined and improved values for the fixed points) that have been in place are:

- the ITS-48, in which minor adjustments were made to the ITS-27, and in 1960 a further modification was made to include the water triple point.
- the IPTS-68 (P = practical), which had a major extension to low temperatures, 13.81 K, being the triple point of hydrogen.
- the ITS-90, in which the range was extended down to 0.65 K, and the type S thermocouple was replaced by extending platinum resistance thermometry up to 962 °C

[7] The Bureau International des Poids et Mesures (BIPM) is the intergovernmental organisation through which Member States act together on matters relating to measurement and measurement standards. It is based at the Pavillon de Breteuil, Sèvres, Paris, France. Website www.bipm.org.

- the PLTS-2000 (the provisional low temperature scale of 2000), covering the range 0.9 mK to 1 K. A specialist temperature scale for researchers into very low temperatures

It is not the purpose of this chapter to go into the details of the various temperature scales. More information about the historical evolution of temperature scales can be found on the BIPM website at: http://www.bipm.org/en/measurement-units/history-si/temperature-scales/

Setting aside the PLTS-2000 which is a specialist temperature scale, the main temperature scale in use around the world is the ITS-90. You can have a thermometer calibrated to the ITS-90 wherever it is implemented from Uruguay, Mexico, US, Germany, Russia, China, Japan, Australia, New Zealand, United Kingdom etc and you will be sure to get a thermometer that reliably measures ITS-90 temperatures with low uncertainties. Given its widespread distribution and its ubiquity I describe here the main features of the ITS-90.

The International Temperature Scale of 1990 (the ITS-90)

The ITS-90 came into effect on 1 January 1990, replacing the IPTS-68, and was the culmination of nearly 100 years of thermometry research. ITS-90 defines procedures by which practical thermometers can be calibrated (using defined fixed points, examples of which are the water triple point and metal freezing points) so that the values of temperature obtained from them are precise and reproducible and approximate to thermodynamic temperatures as closely as possible[8]. It ranges from 0.65 K above absolute zero to the highest achievable temperatures. Its main features are:

- Use of the exquisitely reproducible triple point of water (273.16 K), rather than the freezing point of water (273.15 K), as a defining point.
- Extends down to 0.65 K instead of 13.8 K.
- It is in closer agreement with thermodynamic temperatures, i.e. the defining fixed points have more reliably determined thermodynamic temperatures (these fixed points are given in table 4.1).
- It has a number of overlapping ranges and sub-ranges to allow for the lowest uncertainty calibration of thermometers to be performed.
- There are some special features at lower temperatures, for example it includes the helium vapour pressure scales and an interpolating gas thermometer as defining instruments.
- The range of the platinum resistance thermometer as defining (interpolating) instrument has been extended from 630 °C up to the silver point, 962 °C;
- The type S Pt/10% Rh-Pt thermocouple is no longer a defining (interpolating) instrument of the scale;

[8] http://www.bipm.org/en/committees/cc/cct/publications-cc.html#kelvin-and-temperature-scales.

Table 4.1. The defining fixed points of the ITS-90.

Substance[9]	State[10]	T_{90}/K	t_{90}/ ° C
He	vp	3 to 5	−270 to −268
e-H_2	tp	13.8033	−259.3467
e-H_2 or He	vp or gp	17.035	−256.115
e-H_2 or He	vp or gp	20.27	−252.88
Ne	tp	24.5561	−248.5939
O_2	tp	54.3584	−218.7916
Ar	tp	83.8058	−189.3442
Hg	tp	234.3156	−38.8344
H_2O	tp	273.16	0.01
Ga	mp	302.9146	29.7646
In	fp	429.7485	156.5985
Sn	fp	505.078	231.928
Zn	fp	692.677	419.527
Al	fp	933.473	660.323
Ag	fp	1234.93	961.78
Au	fp	1337.33	1064.18
Cu	fp	1357.77	1084.62

- The range based upon the Planck radiation law begins at the silver point instead of at the gold point, but options exist for using any one of the silver, gold or copper points as reference points for this part of the scale.

The defining fixed points of the ITS-90, their state (e.g. melting or freezing point), ITS-90 temperature in kelvin and degrees Celsius are given in table 4.1; more details are given in the footnotes,[9,10] . It is not appropriate to go into technical details as to how to set up the fixed points, nor how to establish the ITS-90 in this chapter. The interested reader can find more details on the ITS-90 and other aspects of temperature scales in for example the book by Terry Quinn or at the website given in footnote 8.

Temperature scales prior to the introduction of the ITS-90 were succeeded by improved ones on an approximately 20 year timescale. So the ITS-27 gave way to the ITS-48, which in turn was succeeded by the IPTS-68, which was eventually replaced by the ITS-90. The ITS-90 has now been in place for more than 25 years and it seems, at present at least, to remain an effective, fit for purpose scale and is unlikely to be replaced in the near future. However, that does not mean that the scientific discipline of thermometry, and metrology in general, has not been active,

[9] Defined isotopic compositions are recommended for hydrogen, neon and water. All other substances are of natural isotopic composition

[10] The symbols have the following meaning: vp: vapour-pressure point; tp: triple point (temperature at which the solid, liquid and vapour phases are in equilibrium); gp: gas-thermometer point (temperature realized with an interpolating constant-volume gas thermometer); mp, fp: melting point, freezing point (temperature, at a pressure of 101 325 Pa, at which the solid and liquid phases are in equilibrium).

and very significant developments are in train in the coming few years. In the rest of this chapter I focus on current and possible future developments in the science of thermometry.

4.4 The *demise* of defined temperature scales?

In 2018 the conceptual basis of the international system of units (the SI) will be changed. The current ad hoc mixture of classical definitions (e.g. the kilogram and the kelvin), arcane definitions which are no longer used in practice (e.g. the ampere) and fundamental constant definitions (e.g. the metre) will be replaced with a coherent set of definitions based on fixed values of physical constants. In the redefined SI four of the seven SI base units—namely the kilogram, the ampere, the kelvin and the mole—will be redefined. The new definitions will be based on fixed numerical values of the Planck constant (h) for the kilogram, the elementary charge (e) for the ampere, the Boltzmann constant (k) for the kelvin and the Avogadro constant (N_A) for the amount of substance. The present definitions of the metre, the second and the candela will remain unchanged. These changes are well described in the paper by Milton *et al* (2014) and also on the webpage of the BIPM dedicated to the new SI (http://www.bipm.org/en/measurement-units/new-si/).

For the kelvin this means that the current definition which states:

The kelvin, unit of thermodynamic temperature, is the fraction 1/273.16 *of the thermodynamic temperature of the triple point of water.*

(from the SI brochure http://www.bipm.org/en/publications/si-brochure/kelvin.html)

will become after the re-definition in 2018:

The kelvin, symbol K, is the SI unit of thermodynamic temperature; its magnitude is set by fixing the numerical value of the Boltzmann constant to be equal to exactly $1.380\,65X \times 10^{-23}$ *when it is expressed in the SI base unit* $s^{-2}\,m^2 kg\,K^{-1}$, *which is equal to* $J\,K^{-1}$.

In the numerical value, X represents the final digit for the Boltzmann constant which will be agreed shortly before of the redefinition.

The forthcoming redefinition of the SI has driven very significant research around the world in thermometry, first to determine low uncertainty values for the Boltzmann constant to ensure a kelvin redefinition which is consistent with the current definition, and secondly to re-determine the difference between the temperatures obtained from defined temperature scales and those by thermodynamic means, that is a complete redetermination of the quantities $T-T_{90}$ and $T-T_{2000}$[11].

Redetermination of the Boltzmann constant

Very significant research has been undertaken by leading National Measurement Institutes and others around the world to determine very low uncertainty values of the Boltzmann constant. Various physical approaches have been used, including

[11] Where T is the thermodynamic temperature determined from a physical equation, T_{90} and T_{2000} are the temperatures that are obtained from the two defined scales currently in use around the world. In principle (and in practice) the difference between the two values is small because T_{90} is meant to be a close approximation to T at temperatures above 0.65 K and $T2000$ is meant to be a close approximation to T at temperatures from 0.0009 K to 1 K.

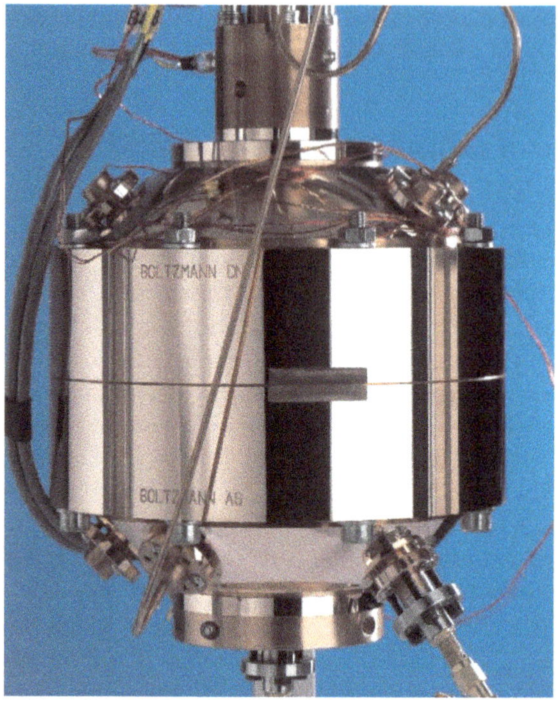

Figure 4.5. The NPL quasi-spherical acoustic resonator, made of diamond-turned copper, in which the Boltzmann constant was measured with uncertainties of less than one part per million, and, the quantity $T-T_{90}$ was determined between 118 K to 303 K, in both cases, with the lowest uncertainties ever attained. Courtesy of NPL.

measuring the speed of sound in gaseous argon and helium, measuring the pressure and dielectric constant of helium, measuring Johnson noise in a resistor and determining the Doppler spectral line width of gas/vapour molecules, all at the triple point of water. These approaches, and the current state of research, are well reviewed in a 2015 edition of the journal *Metrologia* edited by White and Fischer and the interested reader is referred to that journal.

The lowest uncertainty value for the Boltzmann constant has been obtained using the acoustic approach. Here the speed of sound in a very well characterised volume of isotopically known gas is determined at the triple point of water. All these are measured at the state of the art to obtain an overall uncertainty for the Boltzmann constant that is less than one part per million. In figure 4.5 is a photograph of the National Physical Laboratory (NPL) acoustic resonator in which such measurements have been performed.

Redetermination of the difference between the temperatures obtained from defined temperature scales and those by thermodynamic means

Alongside the above described measurements there has been very significant activity to determine the thermodynamic accuracy of the current defined scales. What this means in practice is that the quantities $T-T_{90}$ [12] and $T-T_{2000}$ are being

[12] The quantity $T-T_{90}$, below 400 K is $<\sim 0.01$ K

determined with very low uncertainties using a variety of different thermodynamic temperature measurement techniques.

In the short term these measurements are being performed to enable any users who require thermodynamic temperatures to obtain them from measurements with their thermometers calibrated to the defined scales. In the longer term, such measurement will inform any possible future international temperature scale (the so-called ITS-20xx).

This work is being coordinated through two programmes known as 'Implementing the new kelvin', details of which can be found in volume 374 of the *Philosophical Transactions of the Royal Society* edited by Machin (2016).

What are the implications of these changes for the defined temperature scales?

For all practical purposes, in the immediate short term, these developments should have no significant impact on the practice of thermometry. The current temperature scales, the ITS-90 and the PLTS-2000, will remain in place and be disseminated. However, on another level there will be very *significant* change, particularly in the decades that follow the redefinition.

The redefinition will facilitate and stimulate the drive towards practical primary thermometry. Currently National Measurement Institutes (such as the NPL in the UK) nearly always realise and disseminate the defined temperature scale ITS-90. In the future the situation is likely to become more complex as practical primary thermometry, linked directly to the new kelvin definition, may in some cases yield lower temperature realisation and dissemination uncertainties than the current defined temperature scale.

This emerging and fluid situation will need to be regulated to ensure that there continues to be worldwide uniformity in temperature measurement. This will be undertaken through the very important document known as the *mise en pratique* ('putting into practice') for the definition of the kelvin (the *MeP*-K). This is already in place but more importantly a new version will be produced within 1 year of the unit redefinition, to be known as the *MeP*-K-19. This document will be the mechanism used around the world to ensure that the practice of thermometry remains robust and sound, irrespective of whether the ITS-90, PLTS-2000 or an appropriate thermodynamic method is used to realise and disseminate temperature from National Measurement Institutes to user communities. It will become more important to be clear as to what quantity is being disseminated.

Over time practical primary thermometry approaches will emerge to challenge the defined scales. This is already happening at high temperatures, above 1300 K, where primary spectral radiometry (the measurement of light energy) has similar uncertainties to ITS-90, and at low temperatures below 1 K where electrical noise thermometry may well have lower uncertainties compared to PLTS-2000. It is worth remembering that the defined scales (the ITS-90 and PLTS-2000) are meant to be close approximations to thermodynamic temperature, to help the user measure temperature simply and reproducibly. If thermodynamic temperature realisation and dissemination becomes practical and has uncertainties similar to or lower than the current temperature scale then the need for any defined temperature scale becomes significantly diminished. If widespread practical primary thermometry truly emerges in the decades following the kelvin redefinition and the introduction of the *MeP*-K-19 it is clear that the role of defined temperature scales will be reduced, leading ultimately to their demise, firstly the PLTS-2000 and then probably in a number of decades after the redefinition, the ITS-90. Given the current state of knowledge, the requirements of users and the mechanism of the *MeP*-K-19, it may be that a new temperature scale (ITS-20xx) will be never be needed.

4.5 Summary

I have shown that there has been tremendous scientific progress in both the measurement of temperature and in the understanding of the physical meaning of temperature from the 18th century to the present day. At the opening of the 18th century *de facto* scales, with no physical significance, were the only means of measuring temperature. In the 19th century the true meaning of temperature was established and reliable primary thermometry began to be practised. Temperature measurement in the 20th century was dominated by a succession of *defined* temperature scales which allowed any user to relatively simply determine temperatures that were close to their true thermodynamic values. Unambiguous and significant progress is clearly evident throughout the flow of time.

With the redefinition of the kelvin, and the growing importance of practical primary thermometry, first at the temperature extremes, but increasingly in the middle portion of the temperature scale, it is possible that the ITS-90 could be the last defined temperature scale. Following the redefinition of the kelvin the world-wide uniformity of temperature measurement will be facilitated and ensured through the mechanism of the *mise en pratique* for the definition of the kelvin for the foreseeable future.

Acknowledgements

I thank my employer, the National Physical Laboratory, for giving me time to write this chapter. I thank my wife Catherine for being willing to be dragged around various obscure museums and country houses to hunt down historic thermometers, and engage in arcane conversations about the relative merits of different long extinct temperature scales. I thank the staff of the Museum of History of Science, University of Oxford, for allowing me to inspect their thermometer collection at close quarters, use their library and for the photograph of the Royal Society thermometer in figure 4.2. I thank Dr Paul Lambers, Curator, Universiteitsmuseum, Utrecht for his interest in this work and for providing the photograph used in figure 4.1. Finally, I thank the Institute of Physics History of Physics Group for inviting me to give a talk on this topic, on which this work is based.

References

Cattermole M and Wolfe A F 1987 Horace Darwin's Shop *A history of the Cambridge Scientific Instrument Company 1878–1968* (Boca Raton, FL: CRC Press)

Chang H 2004 *Inventing Temperature: Measurement and Scientific Progress* (Oxford: Oxford University Press)

Fahrenheit D 1724 *Philos. Trans.* **33** 78–84

Gauvin J-F 2012 The Instrument that never was: inventing, manufacturing and branding Reaumur's thermometer during the enlightenment *Ann. Sci.* **69** 515–49

Jameson P R 1914 *The Thermometer and its Family Tree* (Rochester, NY: Taylor Instrument Companies)

Machin G 2016 Towards Implementing the new kelvin *Philos. Trans. R. Soc.* A **374**

Martine G 1740 (1772) An essay towards comparing different thermometers with one another *Essays and observations on the construction and graduation of thermometers* 2nd edn (Edinburgh: Alexander Donaldson)

Middleton W E K 1966 *A History of the Thermometer and its Uses in Meteorology* (Baltimore, MA: John Hopkins)

Milton M J T, Davis R and Fletcher N 2014 Towards a new SI: a review of progress made since 2011 *Metrologia* **51** 21–30

Quinn T 1990 *Temperature* 2nd edn (Amsterdam: Elsevier)

White R and Fischer J 2015 Focus on the Boltzmann Constant *Metrologia* **52**

IOP Publishing

Precise Dimensions
A history of units from 1791-2018
Malcolm Cooper and Jim Grozier

Chapter 5

Kelvin's absolute temperature and its measurement[1]

Hasok Chang

5.1 Thomson's motivations for absolute temperature

William Thomson (1824–1907), better known as Lord Kelvin, made pioneering efforts to create the modern concept of absolute (thermodynamic) temperature, counted in 'degrees Kelvin' in his honor. (I will refer to him as 'Thomson' rather than 'Kelvin' in this chapter, mainly because he was only made Lord Kelvin long after the period of work discussed here.) Aside from its undisputed importance for physics, Thomson's work is particularly interesting because it was driven by an apparently paradoxical impulse. On the one hand, he insisted that the concept of absolute temperature should make no reference whatsoever to any properties of any particular material substances; on the other hand, he demanded that absolute temperature should be a physically measurable quantity, with a clear correspondence to actual thermometer readings. What kind of procedure enabled Thomson to render an entirely abstract concept measurable by concrete physical instruments?

The story of absolute temperature is treated only briefly even in the exhaustive accounts of Thomson's life and work by Silvanus P Thompson (1910), and by Crosbie Smith and Wise (1989), and in the shorter yet insightful biography by Harold I Sharlin (1979). Therefore, the account presented here is primarily based on Thomson's own writings spanning over three decades, making use of various secondary sources for additional insights.[2] The technical derivations presented

[1] This is an expanded version of the presentation given at the Institute of Physics workshop on 'A History of Units from 1791 to 2018' held at the National Physical Laboratory on 17 March 2016. It draws from two previously published sources: Chang (Chang 2004), chapter 4 and Chang and Yi (Chang 2005).

[2] To put Thomson's ideas in the context of the wider development of thermodynamics, see Cardwell (1971, especially pp 239–240, 258–60) and Cardwell (1989), as well as Hutchison (1976). On the broader meanings of 'absolute' measures, see Wise and Smith (1986). Helpful on the technical aspects are Gray (1908, chapter 8, especially pp 123–127, 135–138), and Truesdell (1980, especially sections 11B, 11H, and 9D).

doi:10.1088/978-0-7503-1487-9ch5 © IOP Publishing Ltd 2017

here are interpretations of Thomson's reasoning that should be comprehensible to the technically adept modern reader, while firmly based in Thomson's own general theoretical framework.

In order to understand Thomson's motivations in creating the concept of absolute temperature, it is important to have a sense of the cutting edge of thermal physics at the time, which he was both building on and reacting against. The figure that looms large in that background is Victor Regnault (1810–1878) in Paris, under whom Thomson had learned the ways of experimental physics after his theoretical training through the Cambridge Mathematical Tripos. In his first paper on 'an absolute thermometric scale', published in 1848 in the Proceedings of the Cambridge Philosophical Society, Thomson began by recognizing that the problem of thermometry had received 'as complete a practical solution . . . as can be desired' thanks to the 'very elaborate and refined experimental researches' in recent years, particularly by Regnault. Still, he lamented, 'the *theory* of thermometry is however as yet far from being in so satisfactory a state.' (Thomson [1848] (1882), p 100; emphasis added)

Regnault had consolidated his precision air-thermometry by shrinking from any assumptions that could be questioned, which meant avoiding *all* theories of heat and temperature (Chang 2004, chapter 2). This austere anti-theoretical manner of doing science did not appeal to Thomson, much as he admired Regnault's work. He did appreciate Regnault's impeccable demonstration that the air thermometer was a good instrument to use because it gave highly consistent readings even when its construction was made to vary widely, which was not the case for other common thermometers (for instance, the mercury thermometer and the alcohol thermometer). Even so, he complained:

> Although we have thus a strict principle for constructing a *definite* system for the estimation of temperature, yet as reference is essentially made to a specific body as the standard thermometric substance, we cannot consider that we have arrived at an *absolute* scale, and we can only regard, in strictness, the scale actually adopted as *an arbitrary series of numbered points of reference sufficiently close for the requirements of practical thermometry.* (Thomson [1848] (1882), p 102; emphases original; see also Joule and Thomson [1854] (1882), p 393.)

Thomson wanted to propose a general theoretical principle for thermometry. He first considered the old idea that in a perfect thermometer equal increments of temperature should correspond to equal additions of heat. While not denying the theoretical cogency of this principle in itself, Thomson stated: 'It is however now recognized (from the variation in the specific heats of bodies) as an experimentally demonstrated fact that thermometry under this condition is impossible' (Thomson [1848] 1882, p 100). Here it seems that he was referring to the following problem, which had been widely recognized since around 1800. Since the specific heat of a given object or substance is generally a function of its temperature, the addition of equal amounts of heat to a body actually does not result in the same amount of temperature increase in all parts of the temperature scale.

This would be a mere inconvenience if it were possible to chart the temperature-dependence of specific heat precisely, but that is impossible unless we already have accepted methods of measuring the increments of temperature (and heat). But the method of measuring temperature is precisely what we are trying to devise here. Given that circularity, it is understandable that Thomson declared (Thomson [1848] 1882, p 101): 'we are left without any principle on which to found an absolute thermometric scale.'

5.2 The absolute as the abstract

Thomson was insistent that any fundamental thermometric standard should be 'absolute', not relative to the properties of any particular objects or substances. In his critique of Regnault, his main complaint was that it privileged a particular physical substance, namely air, for no inherent reason. As Wise and Smith show, this idea of absolute measurement permeated Thomson's general 'philosophy of measurement'. His idea of absolute measurement seems to have originated in 1845 when he was pondering about electrical measurements, in fact during his apprenticeship in Regnault's laboratory in Paris. In principle there would have been various types of theoretical relations to serve as bases of absolute measures, but Thomson's preference was to reduce everything to measures of work or energy.

It is interesting to note that Thomson's philosophy of measurement is alive and well at the cutting edge of metrology today. The 20th century trend in metrology had gone against his spirit, defining thermodynamic temperature by reference to the triple point of water, which was given the absolute temperature of 273.16 degrees kelvin at the 10th General Conference on Weights and Measures in 1954. This required, among other things, a specification of what 'water' meant, which was spelled out by 2005 as the 'Vienna Standard Mean Ocean Water', with the ratios of hydrogen and oxygen isotopes specified to five significant figures. But as part of the ongoing reform of the SI system, the kelvin from 2018 will be defined by assigning an exact numerical value to Boltzmann's constant, which has the effect of defining temperature through thermal energy kT (Fellmuth *et al* 2016, pp 2–3). This is much more in keeping with Thomson's orientation, as the new definition would be 'independent of any material substance, technique of realization, and temperature or temperature range' (Fischer *et al* 2007, p 1755). The same impulse can also be felt in the soon-to-be-implemented new definition of the unit of mass, which reduces it down to Planck's constant by means of the Watt Balance (see chapter 9).

For reducing the measure of temperature to the measure of work, the theory of heat engines by Sadi Carnot (1796–1832) gave Thomson precisely what he needed:

> The relation between motive power and heat, as established by Carnot, is such that *quantities of heat*, and *intervals of temperature*, are involved as the sole elements in the expression for the amount of mechanical effect to be obtained through the agency of heat; and since we have, independently, a definite system for the measurement of quantities of heat, we are thus furnished with a measure for intervals according to which absolute differences of temperature may be estimated. (Thomson [1848] 1882, p 102; emphases original)

Carnot's theory provided a theoretical relation between three variables pertaining to an idealized heat-engine: heat, temperature, and work. If heat and work could be measured directly, temperature could be inferred theoretically. Thomson's basic idea in 1848 (to be modified later) was that the interval of one degree of temperature should be defined as that which would result in the production of unit amount of mechanical work in a 'Carnot engine' operating with a unit amount of heat in that temperature interval. In his own words:

> The characteristic property of the scale which I now propose is, that *all degrees have the same value*; that is, that a unit of heat descending from a body *A* at the temperature T° of this scale, to a body *B* at the temperature $(T-1)°$, would give out the same mechanical effect, whatever be the number *T*. This may justly be termed an absolute scale, since its characteristic is quite independent of the physical properties of any specific substance. ((Thomson [1848] 1882, p 104; emphasis added)

This definition is what I will refer to as Thomson's first (concept of) absolute temperature.

It is very important to note that Thomson's sense of 'absolute' here had nothing to do with counting temperature up from absolute zero. In fact, Thomson's 1848 temperature scale did not have a zero point at all. The popular notion of 'absolute zero' that survives into modern times is in fact much older. It can be traced back to Guillaume Amontons (1663–1738), whose work inspired the idea that an objective scale of temperature could be obtained if the zero point were found by extrapolating the observed pressure-temperature relation of air until pressure became zero. What I will call 'Amontons temperature' is quite close to what people commonly mean by 'absolute temperature' nowadays if they have not studied thermodynamic theory carefully. As we shall see later, Thomson later modified his absolute temperature concept to bring it more into line numerically with Amontons temperature; from that point on, the two different senses of 'absolute' (not being related to particular materials, and having an absolute zero) would become forever conflated.

5.3 The operationalization of Thomson's first absolute temperature

Thomson's first absolute temperature constituted an impeccable theoretical definition. But in a way, that was the easy part. Anyone can make up a theoretical definition, but it will not be useful for empirical science unless it can be *operationalized*, i.e., connected to the realm of physical operations. The task of operationalization was made starkly difficult in Thomson's case since he had deliberately fashioned the absolute temperature concept to make sure that any connections whatsoever to any particular objects or materials were severed. How was he going to regain those connections? In 1883 Thomson himself provided a perfect imagery for this situation, taking the viewpoint of a 'scientific traveller roaming over the universe':

> For myself, what seems the shortest and surest way to reach the philosophy of measurement . . . is to cut off all connection with the earth, and think what we

must then do, to make measurements which shall be definitely comparable with those which we now actually make in our terrestrial workshops and laboratories. (quoted in Schaffer 1992, p 42)

A conceptually straightforward scheme for measuring Thomson's first absolute temperature would have been the following: take an object whose temperature we would like to measure; use it as a Carnot heat reservoir and run a Carnot engine between that and another reservoir whose temperature is previously known; measure the amount of mechanical work that is produced, which gives the difference between the two temperatures. The difficulty of realizing that procedure can only be imagined, because no-one has been crazy enough to attempt it. In order to meet the standard of precision in thermometry established by Regnault, the instrument used would have needed to be frighteningly close to the theoretical Carnot engine. That route to the operationalization of absolute temperature was a non-starter.

So Thomson took a conceptual detour. Instead of attempting to measure temperature directly with a thermometer constructed out of an actual heat-engine, he *theorized* about versions of the Carnot engine that were concrete enough to allow the use of certain empirical data in the *description* of its workings. If a measure of absolute temperature could be established in *any* version of the Carnot engine, then it would have universal validity because Thomson's definition of absolute temperature was only based on those features that were shared by all Carnot engines. The key to making a reliable connection between the abstract definition of absolute temperature and actual empirical data was to use a model that was sufficiently concrete, yet still idealized so as to satisfy Carnot's propositions about engine efficiency. Thomson worked out two such models, following Carnot: a system made of water and steam, and a system with only air in it. Here I will only give the details of his water-steam system.[3] An important advantage of this system was that the pressure of 'saturated' steam is a function of temperature only, which simplified the reasoning a great deal as we shall see below. The theoretical model allowed Thomson to compute the heat-work relation from empirical data. As we shall see, the relevant empirical data were certain parameters expressed as functions of temperature measured by an air thermometer. Putting such data into the defining expression for absolute temperature yielded a relation between absolute temperature and air-thermometer temperature, with which he could convert air-thermometer temperature into absolute temperature. Let us now see how this calculation was made.

The net outcome of a complete cycle of operations of the ideal heat-engine (in Carnot's original theory) is the following: a certain amount of work, W, is produced as a certain amount of heat, H, is passed through the system (without being destroyed) from a heat-reservoir at temperature S to a heat-reservoir at temperature T (where $S > T$). We need to evaluate W, which is visually represented by the area enclosed by the quadrilateral $AA_1A_2A_3$ in figure 5.1. In this pressure–volume

[3] For the treatment of the air engine, see Thomson [1849] (1882), pp 127–33.

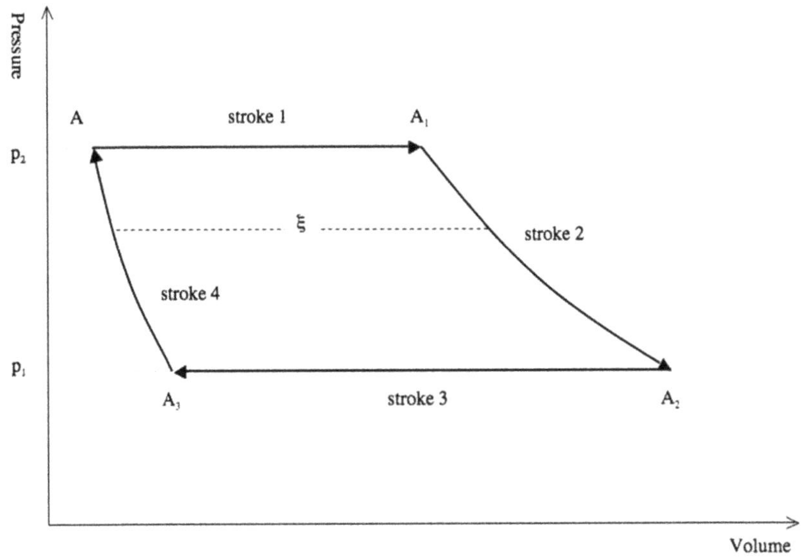

Figure 5.1. Indicator-diagram representation of the working of the ideal steam–water engine, adapted from the figure in Thomson [1849] (1882), p 124.

diagram, the mechanical work produced by an operation is given by the area under the curve that represents it (the integral $\int p\,dv$). In the cycle, the net mechanical work is given by the area of the quadrilateral, which represents the amount of work done *by* the steam–water system in strokes 1 and 2, minus the amount of work done *to* the steam–water system in strokes 3 and 4.

Thomson estimated the area in question actually by performing the integration along the pressure axis, as follows:

$$W = \int_{p_1}^{p_2} \xi \, dp \tag{5.1}$$

where p_1 and p_2 are the pressures in strokes 1 and 3, each constant because the temperature is constant in each stroke. The operative assumption here is that there is a strict correlation between the temperature and pressure of saturated steam, which was generally accepted as an empirical law. What is denoted by ξ is the length of the line parallel to the volume-axis delimited by the curvilinear sides of the quadrilateral. What ξ represent physically is the crucial point (Thomson [1849] (1882), pp 125–6):

> we see that ξ is the difference of the volumes below the piston at corresponding instants of the second and fourth operations, or instants at which the saturated steam and the water in the cylinder have the same pressure p, and, consequently, the same temperature which we may denote by t. Again, throughout the second operation the entire contents of the cylinder possess a greater amount of heat by H units than during the fourth; and, therefore, at any

instant of the second operation there is as much more steam as contains H units of latent heat, than at the corresponding instants of the fourth operation.

Now we must ask how much volume increase results from the production of the amount of steam embodying latent heat H. That volume increment is given as follows:

$$\xi = (1 - \sigma)H/k, \tag{5.2}$$

where k denotes the latent heat per unit volume of steam at a given temperature, and σ is the ratio of the density of steam to the density of water. The formula makes sense as follows: the input of heat H produces H/k liters of steam, from $\sigma H/k$ liters of water; the net increase of volume is given by subtracting the original water volume from the final steam volume.

Substituting that expression into (5.1), the expression for the net work produced in the cycle, we have:

$$W = \int_{p_1}^{p_2} (1 - \sigma)\frac{H}{k}\mathrm{d}p \tag{5.3}$$

Now, because all of the parameters in the above equation except H are measured as a function of air-thermometer temperature t, we can rewrite the integral in terms of t (taking H out as a constant), as follows:

$$W = H\int_{T}^{S} (1 - \sigma)\frac{\mathrm{d}p}{k\,\mathrm{d}t}\mathrm{d}t, \tag{5.4}$$

where S and T are the temperatures of the working substance in the first and the third strokes. According to Thomson's first definition of absolute temperature the difference between those two temperatures on the absolute scale is proportional to W/H, and that ratio can be evaluated by performing the integration in equation (5.4) after putting in the relevant empirical data. Comparing the absolute temperature difference estimated that way with the air-thermometer temperature difference $(S-T)$ gives the conversion factor expressing how many air-thermometer degrees correspond to one degree of absolute temperature interval, at that point in the scale.

Therefore the measurement of absolute temperature by means of the steam-water cycle came down to the measurement of the parameters occurring in the integral in (5.4), namely: the *pressure*, *density*, and *latent heat* of saturated steam as functions of air-thermometer temperature. Detailed measurements of these quantities had been made. Using Regnault's data, Thomson constructed a table with 'a comparison of the proposed scale with that of the air-thermometer, between the limits 0° and 230° of the latter'. Table 5.1 gives some of Thomson's results. Note that the relationship between air–temperature and absolute temperature is not linear; each air-temperature degree 'contained' more and more absolute-temperature degrees as the temperature went down. This absolute scale in fact had no zero-point but stretched to negative infinity.

Table 5.1. Thomson's comparison of air-thermometer temperature and his first absolute temperature. Data from Thomson [1849] (1882), p 139 and p 141.

Air-thermometer temperature	Absolute temperature (first definition)
0 °C	0 °
5	5.660
10	11.243
15	16.751
20	22.184
25	27.545
30	32.834
35	38.053
40	43.201
45	48.280
50	53.291
55	58.234
60	63.112
65	67.925
70	72.676
75	77.367
80	82.000
85	86.579
90	91.104
95	95.577
100	100
150	141.875
200	180.442
231	203.125

Let us now consider whether Thomson at this stage really succeeded in his self-imposed task of measuring absolute temperature. There were three major difficulties. The first one was clearly noted by Thomson himself: the formulae given above require the values of k, the latent heat of steam by volume, but Regnault had only measured the latent heat of steam by weight. Lacking the facility to make the required measurements himself, Thomson converted Regnault's data into what he needed by assuming that steam obeyed the laws of Boyle and Gay-Lussac. He knew that this was at best an approximation, but thought there was reason to believe that it was a sufficiently good approximation for his purposes (Thomson [1848] 1882, pp 104–5).

Secondly, in calculating the amount of mechanical work, the entire analysis was premised on the assumption that the pressure of saturated steam depended only on temperature. That pressure-temperature relation was not something deducible *a priori*, but an empirically obtained generalization, whose rigorous reliability was not beyond doubt. Besides, the use of the pressure–temperature relation of steam amounted to a reliance on an empirical property of a particular substance, just what Thomson wanted to avoid in his theoretical definition of temperature. In his defence, however, we could argue that the strict correlation between pressure and temperature was probably presumed to hold for all liquid-vapor systems, not just for the water-steam system. We should also keep in mind that his use of the pressure-temperature relation was not in the theoretical definition of absolute temperature, but only in its operationalization. Since Carnot's theory gave the assurance that all ideal engines operating at the same temperatures had the same efficiency, calculating the efficiency in any particular system was sufficient to provide a general answer.

Finally, in the theoretical definition itself, absolute temperature was expressed in terms of heat and mechanical effect. We have quoted Thomson above as taking comfort in that 'we have, independently, a definite system for the measurement of quantities of heat', but it is not clear what he had in mind there. The standard laboratory method for measuring quantities of heat was through calorimetry based on the measurement of temperature changes induced in a standard substance (e.g. water), but of course that had to rely on a thermometer. Recall that Thomson's scheme for operationalizing absolute temperature was to express W/H as a function of air-thermometer temperature. A great deal of complexity would have arisen if the measure of H itself depended on the use of the air thermometer (if it had to be kept inside the integral in equation (5.4)). In one place Thomson mentioned using the melting of ice for the purpose of calorimetry, but there were significant difficulties in any actual use of the ice calorimeter (Thomson [1848] 1882, p 106). Still, we could say that in principle heat could be measured by ice calorimetry (or any other method using latent heat), in which case the measure of heat would be reduced to the measure of weight and the latent heat of the particular change-of-state involved. But the last step would end up involving an empirical property of a particular substance, again contrary to Thomson's original intention.

5.4 Thomson's second concept of absolute temperature

How Thomson might have proposed to deal with the above difficulties is an interesting question. However, it is also a hypothetical question. Almost as soon as Thomson advanced his initial concept of absolute temperature, he began the process of abandoning the entire framework in which that concept was couched. This was in large part a consequence of his encounter with James Prescott Joule (1818–1889), the gentleman scientist from a Manchester family of brewers, who is credited with a crucial role in establishing the principle of conservation of energy. When Thomson heard Joule present his idea about the interconvertibility of heat and mechanical work at the 1847 meeting of the British Association for the Advancement of Science in Oxford, he was interested but skeptical. By early 1851, however, Thomson had committed himself to a serious modification of Carnot's theory in the light of Joule's ideas. As a consequence, the entire basis on which he had defined absolute temperature in 1848 had to be changed, because the understanding of the Carnot engine had to be revised fundamentally if heat was no longer considered to be a conserved quantity, and the generation of mechanical effect was seen as the conversion of a part of the heat input into work, rather than a by-product of the movement of heat.[4]

In Thomson's reshaping of the concept of absolute temperature, there were three major steps. The first occurred in the process of reformulating Carnot's theory itself so that it was compatible with energy conservation. Thomson presented his reformulation of Carnot's theory in a series of papers on the 'dynamical theory of heat', starting in March 1851 (Thomson [1851a,1851b,1851c] 1882). The most important part of the new theory, for the purpose of thermometry, was the concept of 'Carnot's function'. Recall that Thomson's original definition of temperature was based on the amount of mechanical effect produced in a Carnot cycle, for a given amount of heat passing through the engine. A crucial factor in such consideration of engine efficiency was what Thomson called 'Carnot's coefficient' or 'Carnot's multiplier', the parameter μ in the following relation:

$$W = \mu H(T_1 - T_2), \tag{5.5}$$

where W is the mechanical work produced in the cycle, H the amount of heat passing through the engine, and T_1 and T_2 are the absolute temperatures of the hot and cold reservoirs.[5] Intuitively, the μ-factor can be seen as a measure of efficiency, indicating how much mechanical effect is produced in an ideal heat engine when a unit amount of heat 'falls' through a unit amount of temperature gap. When Thomson revised Carnot's theory, he preserved a very similar factor, still denoted by μ and called 'Carnot's function'. This was similar to the old Carnot coefficient, but there were important differences (see Thomson [1851a] 1882, section 21, pp 187–8). Because heat was no longer a conserved quantity, H in Equation (5.5) became meaningless, and had to be replaced with something else. Thomson substituted it with the heat *input*, namely the amount of heat absorbed in the first stroke of the cycle (isothermal expansion). In addition, this time μ was defined for a Carnot cycle with an infinitesimal difference between the temperatures of the two heat reservoirs. And μ was now conceived as a function of temperature, while the effect of the 1848 definition of absolute temperature had been to render Carnot's coefficient a constant. With those adjustments, Thomson defined μ through the following work–heat relation parallel to equation (5.5):

$$W = \mu Q \, \mathrm{d}T, \tag{5.6}$$

where $\mathrm{d}T$ is the infinitesimal temperature difference and Q is the heat input.

[4] Although Thomson preserved as much as he could from the old analyses formally, the following claim he made years later seems overstated: 'This paper [of 1848] was wholly founded on Carnot's uncorrected theory… the consequently required corrections… however do not in any way affect the absolute scale for thermometry which forms the subject of the present article.' Thomson did demonstrate that there was a simple numerical conversion formula linking the two definitions of absolute temperature: $T_1 = 100(\log T_2 - \log 273)/(\log 373 - \log 273)$. Note that $T_2 = 0$ puts T_1 at negative infinity, and T_1 is set at 0 when T_2 is 273. See the retrospective note attached to Thomson ([1848] 1882, p 106).

[5] This relation has been extracted from formula (5.7) given in Thomson ([1849] 1882, section 31, p 134), by assuming that μ is a constant, which is a consequence of Thomson's first definition of absolute temperature.

Thomson's second step, simple yet crucial, was to liberalize the theoretical concept of temperature. Carnot's function was related to engine efficiency, to which Thomson still wanted to tie the temperature concept. But he now realized that nothing theoretical actually dictated the exact relation that Carnot's function should bear to temperature; his initial notion of 1848 was too restrictive for no compelling reason. Much more freely, Thomson stated in a paper of 1854 co-authored with Joule: 'Carnot's function . . . *or any arbitrary function of Carnot's function*, may be defined as temperature' (Joule and Thomson [1854] 1882, p 393; emphasis added).

The third step, made possible by the second one, was to find a function of μ that matched existing operational temperature scales reasonably well. In a long footnote attached to a paper of 1854 on thermo-electric currents, Thomson admitted a practical shortcoming of his first definition of absolute temperature, namely that the comparison with air-thermometer temperature showed 'very wide discrepancies, even inconveniently wide between the fixed points of agreement' (as shown in table 1). The most important clue in improving that shortcoming came from Joule, quite unsolicited:

> A more convenient assumption has since been pointed to by Mr Joule's conjecture, that Carnot's function is equal to the mechanical equivalent of the thermal unit divided by the temperature by the air thermometer from its zero of expansion. (Thomson [1854] 1882, p 233, footnote)

What Thomson called 'the temperature by the air thermometer from its zero of expansion' here is Amontons temperature, as I designated it in section 5.2 above. What he called 'Mr Joule's conjecture' can be expressed as follows:

$$\mu = JE/(1 + Et), \tag{5.7}$$

where J is the constant giving the mechanical equivalent of heat, and E is the coefficient of gaseous expansion (Thomson [1851a] 1882, p 199). Thomson took the value of E as 'very nearly equal to 0.003 66 (the centigrade scale of the air-thermometer being referred to)' (Thomson [1849] 1882, p 131). A more intuitive way of writing the formula would be:

$$\mu = J/(273.7 + t_c), \quad \text{or } \mu = J/t_a, \tag{5.8}$$

where t_c is temperature on the centigrade scale, and t_a is Amontons temperature.[6] In this form it is easily recognizable as what Thomson called 'Mayer's hypothesis', in one of its various forms identified by Keith Hutchison (1976, p 279).

Thomson called this proposition Joule's 'conjecture' because he had serious doubts about its rigorous truth. But he thought it was probably approximately true, and therefore capable of serving as a point of departure in finding a concept of absolute temperature closely aligned with practical temperature scales. Thus Thomson used Joule's unverified conjecture as a heuristic device 'pointing to' a

[6] I take the value 273.7 from Joule and Thomson ([1854] 1882), p 394.

new theoretical definition of temperature. In a joint paper with Joule, he stated:

> Carnot's function varies very nearly in the inverse ratio of what has been called 'temperature from the zero of the air-thermometer' [Amontons temperature], . . . and we may *define* temperature simply as the reciprocal of Carnot's function. (Joule and Thomson [1854] 1882, pp 393–4; emphasis added.)

This new idea can be expressed mathematically as follows:

$$\mu = J/T, \tag{5.9}$$

where T denotes absolute temperature. This has basically the same form as equation (5.8), but (5.8) is an empirical hypothesis involving Amontons temperature, and (5.9) is a theoretical definition of absolute temperature.

After giving the above definition, the Joule–Thomson paper added another formulation, which would prove to be much more usable and fruitful:

> If any substance whatever, subjected to a perfectly reversible cycle of operations, takes in heat only in a locality kept at a uniform temperature, and emits heat only in another locality kept at a uniform temperature, the temperatures of these localities are proportional to the quantities of heat taken in or emitted at them in a complete cycle of operations. (Joule and Thomson [1854] 1882, p 394.)

We may write this as follows:

$$T_1/T_2 = Q_1/Q_2, \tag{5.10}$$

where the T's indicate the absolute temperatures of the isothermal processes (strokes 1 and 3 of the Carnot cycle) and the Q's indicate the amounts of heat absorbed or emitted in the respective processes.

How is this alternate formulation justified? Truesdell (1980, p 310) complains that 'Thomson provides no proof' that (5.10) follows from (5.9), and indeed the Joule–Thomson paper itself is not very clear on that point. But it is possible to show instead that (5.9) follows from (5.10), which means that we can take (5.10) as the primary definition. Take a Carnot cycle operating between absolute temperatures T and T' (where $T > T'$), in which the working substance absorbs heat Q in the first stroke and releases Q' in the third stroke ($Q > Q'$). Energy conservation dictates that the net mechanical work produced in that cycle is $J(Q - Q')$, where J is the constant giving the mechanical equivalent of heat, and $(Q - Q')$ is the amount of heat destroyed (converted into mechanical work). Now, using the definition of absolute temperature given in (5.10), we can express the work as follows:

$$W = J(Q - Q') = JQ(1 - T'/T) = JQ(T - T')/T. \tag{5.11}$$

For a cycle in which the temperature difference is infinitesimal, we may write the above equation as follows:

$$W = JQ(\mathrm{d}T/T). \tag{5.12}$$

Now recall the definition of Carnot's function given in equation (5.6), $W = Q\mu dT$. Equating that with (5.12) gives $\mu = J/T$, which is the definition expressed in equation (5.9), so we have the desired result. Definition (5.10) marked a point of closure in Thomson's theoretical work on thermometry, although he would return to the subject many years later. In subsequent discussions we will refer to definitions (5.9) and (5.10) together as Thomson's 'second absolute temperature'.

5.5 The operationalization of the second concept

5.5.1 The ideal gas thermometer as an indicator of absolute temperature

Having clarified the nature of Thomson's second concept of absolute temperature, let us now see how he went about measuring it. Thomson, now in full collaboration with Joule, faced the same basic problem as before: a credible Carnot engine could not be constructed in reality. The operationalization of the second absolute temperature was a long and gradual process, in which a variety of analytical and material methods were tried out by Joule and Thomson, and by later physicists. Most of these methods were based on the assumption, explicit or implicit, that an *ideal* gas thermometer would give the absolute temperature exactly. Then any measure of how much actual gases deviate from the ideal might also give us an indication of how much the temperature indicated by actual gas thermometers deviate from absolute temperature.

The first thing we need to get clear about is why an ideal gas thermometer would indicate Thomson's (second) absolute temperature. The contention to be supported is that an ideal gas expands uniformly with absolute temperature, under fixed pressure (or that its pressure increases uniformly with temperature when the volume is fixed). But a direct experimental test of that idea is impossible, not only because ideal gases do not exist, but also because one would have to know already how to measure absolute temperature so that the gas's behavior can be monitored as a function of absolute temperature. Therefore, any successful argument that an ideal gas indicates absolute temperature has to be made in the realm of theory, not by practical measurement. It is not clear whether Thomson himself made any such argument directly, but at least a plausible reconstruction of his reasoning can be made (see Gray 1908, especially p 125; Zemansky and Dittman 1981, pp 175–7).

The argument is based on an analysis of the isothermal expansion of an ideal gas, such as stroke 1 of the Carnot cycle shown in figure 5.2. In that process the gas absorbs an amount of heat H, while its temperature remains the same. The added heat causes the gas to expand, from initial volume v_0 to final volume v_1, while its pressure decreases from p_0 to p_1; in this expansion the gas also does some mechanical work, because it pushes against an external pressure. The amount of the mechanical work performed, by definition of work, is expressed by the integral $\int p dv$. If we indicate Amontons temperature by t_a as above, then:

$$pv = ct_a, \tag{5.14}$$

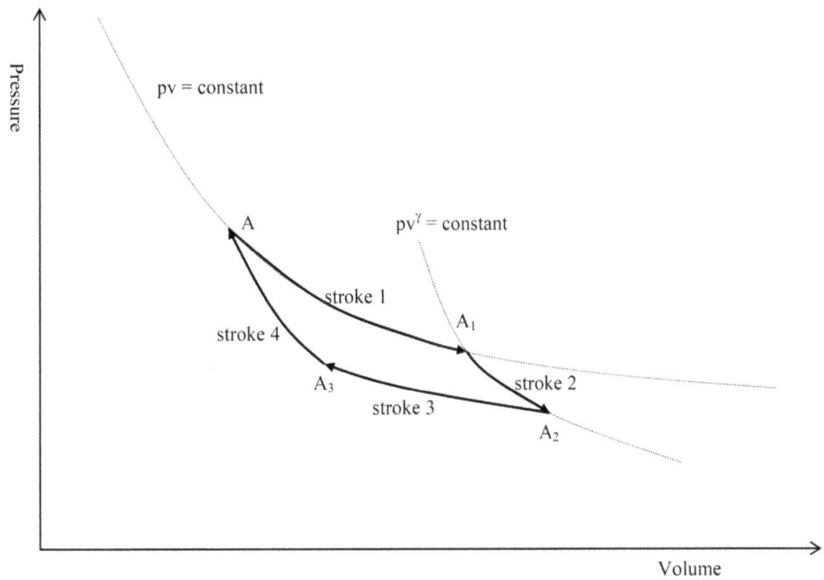

Figure 5.2. Indicator-diagram representation of the working of the ideal gas engine.

where c is a constant specific to the given sample of gas; that is just an expression of the ideal gas law. Putting that into the expression for mechanical work, and writing W_i to indicate the work performed in the isothermal process, we have:

$$W_i = \int_{v_0}^{v_1} p \, \mathrm{d}v = \int_{v_0}^{v_1} \frac{ct_a}{v} \mathrm{d}v. \tag{5.15}$$

Since we are concerned with an isothermal process, the ct_a term can be treated as a constant. So the integration gives:

$$W_i = ct_a \log (v_1/v_0). \tag{5.16}$$

If we ask how W_i varies with respect to t_a, we get:

$$\partial W_i / \partial t_a = c \log (v_1/v_0) = W_i / t_a. \tag{5.17}$$

Now, the variation of W_i with temperature also has a simple relation to Carnot's function (and therefore to Thomson's second absolute temperature). Take equation (5.6), $W = \mu Q \, \mathrm{d}T$, which defines Carnot's function via the net work done in a Carnot cycle in which the temperatures of the first and third strokes (the isothermal processes) differ by an infinitesimal amount, $\mathrm{d}T$. The net work in that infinitesimal cycle is (to the first order) the infinitesimal difference between the work produced in the first stroke and the work consumed in the third stroke. So we may write:

$$W = \frac{\partial W_i}{\partial T} \partial T. \tag{5.18}$$

If we put (5.18) into (5.6), we have:

$$\partial W_i/\partial T = Q\mu = JQ/T. \tag{5.19}$$

In getting the second equality in (5.19) we have invoked the definition of absolute temperature expressed in equation (5.9).

Now compare (5.17) and (5.19). The two equations, one in t_a and the other in T, would have exactly the same form, *if it were the case that* $JQ = W_i$. In other words, the Amontons temperature given by an ideal gas would be interchangeable with absolute temperature, if the condition $JQ = W_i$ were satisfied in an isothermal expansion. But the satisfaction of that condition, according to thermodynamic theory, is the mark of an ideal gas: all of the heat absorbed in an isothermal expansion gets converted into mechanical energy, with nothing going into changes in the internal energy. (And similarly, in adiabatic heating by compression, all the work spent on the ideal gas is converted into heat.) In fact, this condition is none other than Mayer's hypothesis, on the basis of which Joule produced his crucial conjecture, as discussed in the previous section. Therefore, for an ideal gas, t_a and T are interchangeable. This is the best reconstruction of the argument I can give; however, the fact that t_a and T are interchangeable in equations (5.17) and (5.19) does not imply that they are interchangeable in all other respects, so it is not completely demonstrated that the ideal gas thermometer indicates absolute temperature.

5.5.2 Checking the behaviour of actual gases against the ideal

Thomson insisted that Mayer's hypothesis needed to be tested by experiment, and persuaded Joule to collaborate with him on this task (Thomson [1851b] 1882, especially p 211). In the 'porous plug experiment' Joule and Thomson investigated the passage of a gas through a narrow opening. This ought to be an isothermal process for an ideal gas, but not for an actual gas. In an ideal gas lacking intermolecular forces, a free expansion would not require any energy to be spent. But an actual gas has cohesion, so expanding it would require some energy (as it takes work to stretch a spring); this would cool down the gas because the necessary energy would have to be taken from the thermal energy of the gas itself, in the absence of an external energy source. The magnitude of this cooling effect would provide a measure of how much the actual gas deviates from the ideal. The basic scheme of Joule and Thomson's experiment consisted in forcing a continuous stream of gas through two (spiral) pipes connected to each other through a very small orifice.

Because it was difficult to measure precisely the temperature of the gas exiting from the orifice, Joule and Thomson instead measured the amount of heat that was required in bringing the gas back to its original temperature after its passage. From this amount of heat and the specific heat of the gas, the temperature at which the gas had exited from the orifice was inferred. The results of the experiment enabled Thomson and Joule to derive an equation for the volume and pressure of actual gases as a function of absolute temperature, showing how they deviated from the ideal gas law. Their 'complete solution' was the following (Joule and Thomson [1862] 1882, p 430):

$$v = \frac{Ct}{p} - \frac{1}{3}AJK\left(\frac{273.7}{t}\right)^2. \tag{5.20}$$

This equation of state for actual gases expresses t, 'the temperature according to the absolute thermodynamic system of thermometry', in terms of other parameters, all of which are presumably measurable: v is the volume of a given body of gas; p is its pressure; C is a parameter 'independent of both pressure and temperature'; A seems to be a constant that is characteristic of each type of gas; J is the mechanical equivalent of heat; and K is the specific heat per unit mass of the gas under constant pressure. So equation (5.20) in principle indicates a straightforward way of measuring absolute temperature. The second term on the right-hand side gives the measure of how much the gas deviates from the ideal; without it, equation (5.20) would simply reduce to the ideal gas law, which would mean that the conventional gas thermometer correctly indicates the 'temperature according to the absolute thermodynamic system of thermometry'. The derivation of equation (5.20) was a complex theoretical and experimental matter on which Joule and Thomson spent a decade (see Chang and Yi 2005, section 5.5, for details). Here I only discuss its consequences.

Now it seemed that Thomson and Joule had finally succeeded in reducing absolute temperature to measurable quantities, and they in fact proceeded to compute some numerical values for the deviation of the air thermometer from the absolute scale. The results (table 5.2) were quite reassuring for the air thermometer: although the air-absolute discrepancy increased steadily as the temperature increased, it was estimated to be only about 0.4 °C at around 300 °C for Regnault's standard air thermometer.

5.6 Iterative operationalization

5.6.1 The problem of circularity

Joule and Thomson were quite understandably pleased with the above results. However, there remained a serious quandary. The cooling effect in the Joule–Thomson experiment described above, as well as the constants J and K in equation (5.20), were all calculated on the basis of temperature values measured by ordinary thermometers. That is to say, the correction of the air thermometer was done by the measurement of the cooling effect in the 'porous plug' experiment measured by Joule and Thomson, but the latter measurement was given by the air thermometer itself (actually, by the mercury thermometer underwritten by the air thermometer). In that situation, how can we be assured that the *correction* was *correct*? Note that the situation here is fundamentally different from that of Thomson's scheme for operationalizing his first definition of absolute temperature. In that case, the use of empirical data taken with the air thermometer did not pose a problem because Thomson was seeking an explicit correlation of absolute temperature and air-thermometer temperature.

Table 5.2. Joule and Thomson's comparison of absolute temperature (second definition) and air-thermometer temperature. Data from Joule and Thomson [1854] 1882, pp 395–396.

Absolute temperature, minus 273.7°	Air-thermometer temperature
0	0
20	20 + 0.0298
40	40 + 0.0403
60	60 + 0.0366
80	80 + 0.0223
100	100
120	120 − 0.0284
140	140 − 0.0615
160	160 − 0.0983
180	180 − 0.1382
200	200 − 0.1796
220	220 − 0.2232
240	240 − 0.2663
260	260 − 0.3141
280	280 − 0.3610
300	300 − 0.4085

Thomson recognized this problem clearly many years later, in his entry on 'heat' in the 9th edition of the *Encyclopaedia Britannica* (Thomson 1880, p 49, section 55), where he confessed with great insight: 'we have no right to measure these [Joule–Thomson] heating and cooling effects on any scale of temperature, as we have not yet formed a thermometric scale....' He indicated how the problem could be avoided, in principle: 'Now, instead of reckoning on any thermometric scale the cooling effect or the heating effect of passage through the plug, we have to measure the quantity of work (δw) required to annul it.' But he admitted that 'the experiments as actually made by Joule and Thomson simply gave the cooling effects and heating effects shown by mercury thermometers.' The justification that he produced at the end of this remarkable discourse is disappointing:

> The very thermometers that were used [in the Joule–Thomson experiment] had been used by Joule in his original experiments determining the dynamical equivalent of heat [J], and again in his later experiments by which for the first time the specific heat of air at constant pressure [K] was measured with sufficient accuracy for our present purpose. Hence by putting together different experiments which had actually been made with those thermometers of Joule's, the operation of measuring δw, at all events for the case of air, was virtually completed. Thus according to our present view the mercury thermometers are merely used as a step in aid of the measurement of δw, and their scales may be utterly arbitrary. . . .

What Thomson claims here is that the temperature measurements are merely ways of getting at the value of the quantity δw, and the final result is independent of the particular method by which it is obtained. This claim is hollow, unless it happens to be the case that the resulting empirical formula for δw is not a function of mercury-temperature at all. But δw *is* a function of mercury-temperature (t) in general, according to Joule and Thomson's own results. The empirical formula derived from their experiments was the following:

$$-\mathrm{d}\vartheta/\mathrm{d}p = A(273.7/t)^2, \tag{5.21}$$

where ϑ is the temperature change that the gas undergoes in passing through the narrow opening, and A is a constant whose value depends on the nature of the gas (A was determined to be 0.92 air, and 4.64 for carbon dioxide); δw is given by multiplying ϑ by K, the specific heat of the gas, and J, the mechanical equivalent of heat (Joule and Thomson [1862] 1882, pp 428–9). I do not see how it can be argued that δw would in general have no dependence on t. The same point can be seen even more clearly if we take the view that certain errors are introduced into the measured values of ϑ, K and J, if those values are obtained on the basis of the assumption that the mercury thermometer readings indicate the absolute temperature. Thomson's claim amounts to insisting *a priori* that all such errors cancel each other out when the three quantities are multiplied together to produce the final result. That is possible in particular cases, but by no means guaranteed.

This, however, seems to be where Thomson left the problem. In the corpus of his work after the *Britannica* article and a couple of related papers published around the same time, I have not found any further contributions to the measurement of absolute temperature. He was apparently quite satisfied with the theoretical understanding of absolute temperature that he had been able to secure in the framework of a fully developed theory of thermodynamics, and in practical terms he was happy with the old Joule–Thomson empirical results that seemed to give a sufficient indication that the deviations of gas-thermometer temperature from his second absolute temperature were quite small.

5.6.2 An iterative solution to the circularity problem

Fortunately, a much more satisfying understanding of the problem of operationalizing absolute temperature was to emerge within a decade, apparently starting with the work of Hugh Longbourne Callendar (1863–1930), English physicist and engineer who made important observations on the properties of steam and crucial contributions to electric-resistance thermometry. My discussion will rely on the exposition given by Henri-Louis Le Chatelier (Le Chatelier and Boudouard 1901, pp 23–6), which is much more helpful than Callendar's own (1887, p 179). The Callendar–Le Chatelier operationalization of absolute temperature can be understood as an instance of the process of 'epistemic iteration' (see Chang 2004, chapter 5).

The starting point of epistemic iteration is the affirmation of a certain system of knowledge, which does not have an ultimate justification and may need to be changed later for various reasons. The initial assumption for Callendar was that air-thermometer temperature and absolute temperature values were very close to each other. We start by writing the law governing the thermal behavior of actual gases as follows:

$$pv = (1 - \phi)RT, \tag{5.22}$$

where R is a constant, T is absolute temperature, and ϕ is an as-yet unknown function of T and p. The factor ϕ is what makes equation (5.22) different from the ideal gas law, and it is a different function for each type of gas; it is *presumed* to be small in magnitude, which amounts to an assumption that actual gases roughly obey the ideal gas law. Such an assumption is not testable (or even fully meaningful) at that stage, since T is not operationalized yet; however, it may be vindicated if the correction process is in the end successful, or discarded as implausible if the correction process cannot be made to work.

The next step is to estimate ϕ, which is done by means of the results of the Joule–Thomson experiment discussed in section 5.2 above. Le Chatelier gives the following empirical result, calculated from the data obtained in experiments with atmospheric air:

$$\phi = 0.001173\frac{p}{p_o}\left(\frac{T_o}{T}\right)^3, \tag{5.23}$$

where p_o is the standard atmospheric pressure and T_o is the absolute temperature of melting ice. I have presented the derivation of this result in some detail elsewhere (Chang 2004, chapter 4), but one important point can be gathered from merely inspecting the final outcome. Equation (5.23) is supposed to be an empirical result, but it expresses ϕ as a function of absolute temperature T, not as a function of t_a, (Amontons) temperature measured by an ordinary thermometer in the Joule–Thomson experiment. What happens in this derivation is a *deliberate conflation* of absolute temperature and air-temperature (or mercury-temperature), as Callendar and Le Chatelier take the empirical Joule–Thomson formula expressed in t_a and simply substitute it into theoretical formulas expressed in T, letting t_a stand in for T. This is allowed, as an approximation, on the assumption that T and t_a are roughly equal because ϕ is very small.

Unlike Thomson, Le Chatelier was very clear that equation (5.23) did not give the final correction (Le Chatelier and Boudouard 1901, 5.25): 'This is still an approximate result, for we have depended upon the experiments of Joule and Thomson and on the law of adiabatic expansion.' Here Le Chatelier was also acknowledging the fact that in the derivation of (5.23) he had helped himself to the adiabatic gas law, knowing that it was not known to be exactly true but assuming that it was approximately true. A further round of corrections could be made with the help of the correction indicated in (5.23). This would involve re-calibrating the air thermometer, according to the law of expansion that is obtained by inserting (5.23) into (5.22); recall that the air thermometer was initially calibrated on the basis of the assumption that the expansion of air was exactly regular ($\phi = 0$). With the re-calibration of the air thermometer, one could either do the Joule–Thomson measurements again or re-analyze the old data. Either way, the refined version of the experiment would yield a more refined estimate of ϕ, giving an updated version of (5.23). This process could be repeated as often as desired. A similar assessment of the situation was given twenty years later by A L Day and R B Sosman, showing the most succinct conceptual clarity on the matter that I have seen:

> It is important at this point to recall that our initial measurements with the gas-thermometer tell us nothing about whether the gas in question obeys the law $pv = k\theta$ or not. Only measurements of the energy-relations of the gas can give us that information. But since such measurements involve the measurement of temperature, *it is evident that the realisation of the temperature scale is logically a process of successive approximations.* (Day and Sosman 1922, p 837; emphasis added.)

However, it seems that in practice no-one was worried enough to enter into second-round corrections or beyond. Callendar calculated the first-round corrections on air up to 1000 °C; although the corrections got larger with increasing temperature, they turned out to be only 0.62° at 1000 °C for the constant-volume air thermometer, and 1.19° for the constant-pressure air thermometer. It was seen that the corrections would grow rapidly beyond that point, but that was not so much of a practical concern since 1000 °C was about the very limit at which any gas thermometers could be made to function at all in any case.[7] Le Chatelier was happy to declare:

> The deviations of the air-thermometer at high temperatures are thus very slight if concordance is established at 0° and 100°; we shall not have to occupy ourselves further with the differences between the indications of the thermodynamic thermometer and those of the gas-thermometer. (Le Chatelier and Boudouard 1901, p 26)

One only needed to avoid gases like carbon dioxide, for which the corrections were significantly larger. Day and Sosman gave a similar view (1922, p 837): 'Practically, the first approximation is sufficient, so nearly do the gases commonly used in gas-thermometers conform to the "ideal" behaviour expressed in the law $pv = k\theta$.'

This is a pleasing result, but we must also keep in mind that the smallness of the first-round correction is hardly the end of the story. First of all, we would need to see whether the corrections actually continue to get smaller in such a way as to result in convergence. So the only thing we can do is to carry on with the iteration until we are pragmatically satisfied that a convergence seems destined to happen. Moreover, if we are to respect Thomson's original aim of taking the definition of temperature away from particular substances, various gas thermometers need to converge not only each in itself, but all of them with each other. Only then could we have a perfect

[7] See Callendar (1887, p 179). According to Day and Sosman (1922, p 859), up to that time only four attempts had been made to reach 1000 °C with gas thermometers.

match between the single-valued image of absolute temperature and the operational absolute temperature measured by a collection of gas thermometers. It is perhaps plausible to reject some particular gases as legitimate thermometric fluids if there are particular reasons that should disqualify them, but at least some degree of generality would need to be preserved.

5.6.3 Implications of the iterative solution

Seeing the 'correction'of actual thermometers as an iterative process clarifies some issues that have been left obscure in my analysis so far. The clarification stems from the realization that in an iterative process, point-by-point justification of each and every step is neither possible nor necessary; what matters is that each stage leads on to the next one with some improvement. The point here is not only that slightly incorrect information fed into an iterative process may well be corrected. The question of correctness does not even apply, unless and until the iterative process produces a successful outcome, which we then take as the correct answer. Therefore, it makes sense to relax the sort of demand for justification that can lead us to seek illusory rigor.

There are several aspects of this relaxation. (1) First of all, one's exact starting point may not be important. In the case of absolute temperature, assuming the ideal gas law to be approximately true happened to hit the nail nearly on the head, but the iterative correction process could also have started from other initial approximations and reached similar final results. (2) Just as different starting points may lead toward the same conclusion, different paths of reasoning may do so as well. Thomson himself proposed various methods of operationalizing absolute temperature, though only one was pursued sufficiently so it is difficult to know whether the same outcome would have been reached through his other strategies. But the Joule–Thomson experiment was not the only possible way to obtain the desired results. (3) Some looseness can also be allowed in the process of reasoning adopted beyond the initial starting-point. Thomson was able to make certain shortcuts and apparently unwarranted approximations in his various derivations without much of a tangible consequence. Similarly, Le Chatelier helped himself to the adiabatic gas law, knowing full well that it was not guaranteed to be exactly right. (4) Empirical data that may not be exactly right can also be used legitimately. Therefore, Thomson's defence of the use of Joule's mercury thermometer in the Joule–Thomson experiment was not only invalid, but also unnecessary. A recognition of the nature of the iterative process would have spared Thomson from an illusory problem and a pseudo-solution to it.

One more important issue remains to be clarified. In the process of operationalizing an abstract concept, what exactly do we aim for, and what exactly do we get? The hoped-for outcome is an agreement between the concrete image of the abstract concept and the actual operations that we adopt for an empirical engagement with the concept (including its measurement). That is the correspondence that makes the most sense to consider, not the complacently imagined correspondence between theory and experience, or theory and 'reality'. With an iterative process we do not expect ever to have an exact agreement between the operational image and the actual operations, but we hope for a gradual convergence between them. Such convergence would be a considerable achievement, especially if it could be achieved with a high degree of quantitative precision.

This convergence provides a basis for a workable notion of accuracy. We can say that we have an accurate method of measurement, if we have good convergence. How about truth? Can we ever say whether we have obtained the true values of an abstract concept like absolute temperature? The question of truth only makes sense if there is an objectively determinate value of the concept in each physical situation. If we have a convergent operationalization, we could consider the limits of convergence as the 'real' values; then we can use these values as the criteria by which we judge whether *other* proposed operationalizations produce true values. But we must keep firmly in mind that the existence of such 'real values' hinges on the success of the iterative procedure, and the successful operationalization is constitutive of the 'reality'. If we want to please ourselves by saying that we can approach true values by iterative operationalization, we also have to remember that this truth is a destination that is only created by the approach itself.

References

Callendar H L 1887 On the practical measurement of temperature: Experiments made at the Cavendish laboratory Cambridge *Philos. Trans. R. Soc. London* **A178** 161–230

Cardwell D S L 1971 *From Watt to Clausius* (Ithaca: Cornell University Press)

Cardwell D S L 1989 *James Joule: A Biography* (Manchester: Manchester University Press)

Chang H 2004 *Inventing Temperature: Measurement and Scientific Progress* (New York: Oxford University Press)

Chang H and Yi S W 2005 The absolute and its measurement: William Thomson on temperature *Ann. Sci.* **62** 281–308

Day A L and Sosman R B 1922 Temperature realisation of absolute scale of ed R Glazebrook *A Dictionary of Applied Physics vol 1 Mechanics Engineering Heat* (London: Macmillan) 836–71

Fellmuth B *et al* 2016 The Kelvin redefinition and its *mise en pratique Philos. Trans. R. Soc.* A **374** 20150037

Fischer J *et al* 2007 Preparative steps towards the new definition of the Kelvin in terms of the Boltzmann constant *Int. J. Thermophys.* **28** 1753–65

Gray A 1908 *Lord Kelvin: An Account of His Scientific Life and Work* (London: J M Dent and Co.)

Hutchison K 1976 Mayer's hypothesis: A study of the early years of thermodynamics *Centaurus* **20** 279–304

Joule J P and Thomson W 1854 1882 On the thermal effects of fluids in motion part 2 in Thomson (1882) 357–400 Originally published in 1854 in the *Philos. Trans. R. Soc.* **144** 321–64

Joule J P and Thomson W 1862 1882 On the thermal effects of fluids in Motion part 4 in Thomson (1882) 415–31 Originally published in 1862 in the *Philos. Trans. R. Soc.* **152** 579–89

Le Chatelier H and Boudouard O 1901 *High–Temperature Measurements* ed G K Burgess (New York: Wiley)

Schaffer S 1992 Late Victorian metrology and its instrumentation: A manufactory of Ohms ed R Bud and S E Cozzens *Invisible Connections: Instruments Institutions and Science* (Bellingham, WA: Spie Optical Engineering Press) 23–56

Sharlin H I 1979 *Lord Kelvin: The Dynamic Victorian*in collaboration with ed T Sharlin (University Park and London: The Pennsylvania State University Press)

Smith C and Wise M N 1989 *Energy and Empire: A Biographical Study of Lord Kelvin* (Cambridge: Cambridge University Press)

Thompson S P 1910 *The Life of William Thomson Baron Kelvin of Largs* 2 vols (London: Macmillan)

Thomson W 1848, 1882 On an absolute thermometric scale founded on Carnot's theory of the motive power of heat and calculated from Regnault's observations in Thomson (1882) 100–6 Originally published in 1848 in the Proceedings of the Cambridge Philosophical Society 1 66–71; also in the *Philosop. Magazine* (3rd series) **33** 313–7

Thomson W 1849, 1882 An account of Carnot's theory of the motive power of heat; with numerical results deduced from Regnault's experiments on steam in Thomson (1882) 113–55 Originally published in 1849 in the *Trans. R. Soc. Edinburgh* **16** 541–74

Thomson W 1851a, 1882 On the dynamical theory of heat with numerical results deduced from Mr Joule's equivalent of a thermal unit and M Regnault's Observations on steam in Thomson (1882) 174–210 (presented as Parts 1–3 of Art 48 under this title) Originally published in 1851 in the Transactions of the Royal Society of Edinburgh; also in 1852 in the Philosophical Magazine (4th series) **4**

Thomson W 1851b, 1882 On a method of discovering experimentally the relation between the mechanical work spent and the heat produced by the compression of a gaseous fluid in Thomson (1882) 210–22 (presented as Part 4 of Art 48) Originally published in 1851 in the *Trans. R. Soc. Edinburgh* **20** 2

Thomson W 1851c, 1882 On the quantities of mechanical energy contained in a fluid in different states as to temperature and density in Thomson (1882) 222–32 (presented as Part 5 of Art 48) Originally published in 1851 in the *Trans. R. Soc. Edinburgh* **20** 3

Thomson W 1854, 1882 Thermo–electric currents in thomson (1882) 232–291 (presented as Part 6 of Art 48) Originally published in 1854 in the *Trans. R. Soc. Edinburgh* **21** 1

Thomson W 1880 *Elasticity and Heat: Being Articles Contributed to the Encyclopaedia Britannica* (Edinburgh: Adam and Charles Black)

Thomson W 1882 *Mathematical and Physical Papers* vol 1 (Cambridge: Cambridge University Press)

Truesdell C 1980 *The Tragicomical History of Thermodynamics*1822–1854 (New York: Springer)

Wise M N and Smith C 1986 Measurement work and industry in Lord Kelvin's Britain *Historical Studies in the Physical Sciences* **17** 147–73

Zemansky M W and Dittman R H 1981 *Heat and Thermodynamics* 6th ed (New York: McGraw-Hill)

IOP Publishing

Precise Dimensions
A history of units from 1791-2018
Malcolm Cooper and Jim Grozier

Chapter 6

A brief history of the unit of chemical amount

Juris Meija

Humans have been making measurements of chemical nature for a long time. The use of gold and silver as a currency has created the need for analyzing the purity of the precious metals. Amarna letters, written on clay tablets over 3000 years ago, detail the complaint of the king of Babylon to Amenhotep III (Rainey 2015):

> As for the forty minas of gold that they brought, when I cast it into the kiln, for sure only ten minas came forth

Thus, gold was tested for its purity by the fire-assay, a process also known as cupellation. Roman emperor Nero required that taxes should be paid with gold that was tested using such method. This process is described in full detail for the first time only in the 16th century in one of the first textbooks of quantitative analytical chemistry—*De re metallica* by Georgius Agricola (1556)[1].

In his book *On Stones* (circa 300 BC), Greek philosopher Theophrastus describes a smooth stone that imparts a colour of metal alloys when rubbed (Wälchli 1981). A few centuries later, Pliny the Elder writes in his *Natural History* that with this method one can 'tell in a moment the proportion of gold there is in it, how much silver, or how much copper; [...] their accuracy being so marvelous that they are never mistaken' (Pliny the Elder). Because the colour of precious alloys depends largely on the gold content, their colour can be compared to samples of known purity. This early form of colorimetric touchstone comparison method was used by Romans and Egyptians alike and this practice continues to this day.

[1] As an interesting historical note, this book was first translated into English in 1912 by Lou Henry Hoover and her husband Herbert Hoover who later became the 31st President of the United States of America.

doi:10.1088/978-0-7503-1487-9ch6

© IOP Publishing Ltd 2017

6.1 Comparative measurements

In 1663, Robert Boyle noted that acids and bases can destroy and restore the blue colour of some plant dyes such as the syrup of violets. This enabled him to 'guess at the strength of the liquors, thus examined, by the quantity of them; which is sufficient to destroy, or restore the blue colour of our tincture' (Boyle 1725). In 1699, Wilhelm Homberg, who had worked with Boyle in his laboratory, reported quantitative measurements on neutralization of acids and bases. He measured the strength of acids by neutralizing them with potassium carbonate and then weighing the dried residue (Homberg 1699). Similar experiments were done in 1729 by Claude Francois Geoffroy who determined the strength of vinegar by adding a powder of potassium carbonate until no further effervescence took place. These three early studies seem trivial to us but they paved a way for entirely new kinds of chemical measurements which enable determining the amounts of several substances by chemical means. Furthermore, the analysis methods of Boyle, Homberg, and Geoffroy differ from those described by Pliny the Elder and Agricola because they enabled chemists, for the first time, to chemically *compare* the quantities of various substances. Before the 18th century, measurements of different chemical quantities were seen as unrelated. One can determine the mass of copper and gold in a coin but a direct comparison of these two values does not seem to be of much value. However, comparing the amount of vinegar with that of potash enables us to express the strength of vinegar in terms of a universal unit, for example, a teaspoon of potash. The idea of comparing amounts of chemical substances thus emerged in the 17th and 18th century. In 1767, British chemist Henry Cavendish described a fixed mass of potassium carbonate as 'equivalent' to a fixed mass of calcium oxide if they would both saturate (neutralize) equal amounts of acid.

6.2 Quantitative measurements

In the late 18th century, French chemist Antoine-Laurent Lavoisier revolutionized chemistry through his experiments and ideas. In his seminal book *Traité élémentaire de Chimie* (1789), Lavoisier put forward that both the mass and matter is conserved in chemical reactions (Lavoisier 1790, Smartt Bel 2005):

> We may lay it down as an incontestible axiom, that, in all the operations of art and nature, nothing is created; an equal quantity of matter exists both before and after the experiment; the quality and quantity of the elements remain precisely the same; and nothing takes place beyond changes and modifications in the combination of these elements. Upon this principle the whole art of performing chemical experiments depends.

Lavoisier was certainly not the first to recognize this fundamental law, but he is credited as its discoverer because he applied the permanency of mass to chemistry and all chemical processes alike. As Thomas Kuhn has noted (Kuhn 1961), Lavoisier's contemporaries felt that his theory of combustion deprived chemistry of one of its principal traditional functions—the explanation of the qualitative properties of bodies. In addition to introducing chemists to quantitative thought, a striking feature of the *Traité* was the first modern definition of chemical elements. These advances prompted chemists to further study the relationship between the masses of various substances that combine with one another, known as the equivalent or combining weights. In the 1790s, his fellow countryman Jeremias Benjamin Richter called this study 'stoichiometry' or 'the art of measuring chemical elements'. Richter found that the mass ratio of the compounds consumed in a chemical reaction was always the same: when two substances combine with a third in a certain proportion, they will combine with each other in the same proportion. Today we know this as the law of reciprocal proportions. The same conclusion was reached by the French chemist Joseph-Louis Proust who observed that substances always contain their component elements in fixed mass ratio regardless of method of preparation. Richter compiled a list of values for various common chemicals which were 'equivalent' to one another. For example, 1000 parts of hydrochloric acid were equivalent to 858, 1107, and 3099 parts of magnesium, calcium, and barium oxides, respectively. Likewise, 1000 parts of sulphuric acid were equivalent to 616, 796, and 2226 parts of magnesium, calcium, and barium oxides, respectively. By comparing the numerical values in these two series, Richter thus found a way not only to compare amounts of substances that react directly with one another, such as acids and bases, but also a way to compare amounts of substances that do not react with one another, such as sulphuric and hydrochloric acids. Richter showed that 1394 parts of sulphuric acid and 1000 parts of hydrochloric acid were distinct yet equivalent quantities[2]. In other words, 1394 parts of sulphuric acid will neutralize the same amount of other bases as 1000 parts of hydrochloric acid.

In contrast to the view that chemical substances unite in definite proportions, some prominent chemists of the time thought that in most cases substances could combine chemically in an infinite number of ratios varying continuously between certain limits (Hartog 1894), a notion that was soon disproved by the precise experimental work of the Swedish chemist Jöns Jacob Berzelius (1779–1848).

The general aptitude for quantitative work by chemists was in its infancy in the late 18th century. In 1786, Immanuel Kant famously denied chemistry the status of 'proper' science largely because of the lack of application of mathematics in its teachings (McNulty 2014). To Kant, there could be only as much proper science as there is mathematics therein. Perhaps this explains the rudimentary level of Richter's explanations in his *Stoichiometry* where he famously explains how to add two numbers '19 + 424 means that we add 19 to 424' (Szabadváry 1966). A century later, similar sentiments were raised by Sir Edward Frankland, one of the leading chemists of his time, in a letter to James Joseph Sylvester, a leading mathematician of his time: 'I am convinced that the future progress of chemistry, as an exact science, depends very much indeed upon the alliance of mathematics' (Frankland 1878).

Richter spent a significant amount of time trying to explain the numerical patterns in the values of equivalent weights. For example, he noted that the

[2] Equivalent weights are often wrongly attributed to C F Wenzel (1777). See *A Short History of Chemistry* By James Riddick Partington.

equivalent weights of magnesium, calcium, and barium oxides (in a scale HCl = 1000) can be represented in a mathematical progression $a + b$, $a + 3b$, and $a + 19b$ where $a = 734$ and $b = 124.5$[3]. Richter was, however, not alone in trying to find patterns in the equivalent weights of the various elements. In fact, throughout a large part of the 19th century chemists would search for meaning in the atomic weights in order to find universal relationships between these values. Most notably, this led the English chemist William Prout (1785–1850) to suggest in 1815 that atomic weights of all the elements are exact multiples of hydrogen (Prout 1815).

In 1808 French chemist Joseph Louis Gay-Lussac published his observations on the combining volumes of gases, namely that they react with one another in very simple proportions of their volumes. Moreover, if the product is also a gas, its volume too is in simple proportion of the reactants. Armed with the results of Gay-Lussac, Amedeo Avogadro postulated in 1811 that gases at the same temperature and pressure contain the same number of molecules. Avogadro saw his postulate as a practical way to establish accurate relative masses of atoms based on the density of gases.

6.3 The mass unit of the chemist: the gram-molecule

One of the fundamental laws of chemistry is that of discontinuity. The discontinuous variation according to the law of multiple proportions in the composition of various substances made from the same atoms becomes immediately clear: is arises solely due to the condition that the molecule constituting a compound contains necessarily a whole number of atoms of each kind. Indeed, John Dalton (1803) noted that 'when elements combine, they do so in a ratio of small whole numbers'. This was a culmination of the work of scientists such as Wenzel, Richter, and Proust. Here Dalton meant not the mass but rather the number of atoms. Thus began the paradigm shift away from inertial descriptors of chemical processes (such as the mass or the volume of the reacting substances) to the particulate descriptors (such as the number of atoms or derived quantities directly proportional to them). After all, not everything depends on the mass of substance. Fill equal masses of two gases in two identical cylinders and they will exert different pressure. Combine equal masses of sulphuric acid and sodium hydroxide and the resulting mixture will be far from neutral.

Chemists accepted that the amount of substance is characterized by the number of elementary entities and not their mass. Although there was no means to ascertain this number directly, Dalton put forward the idea that chemical elements are made of atoms of differing mass. Combined with the idea that elements combine with one another in a ratio of small whole numbers, Dalton could deduce the relative masses of the various atoms using hydrogen as the unit (Dalton 1808). For example, oxygen and hydrogen unite in a mass ratio of 8:1 to form water. If we assume that water contains 1 atom of hydrogen and 1 atom of oxygen, as Dalton did, then the 8:1

[3] To Richter, the unfilled sequences in the series from $a+3b$ … $a+19b$ represented still undiscovered elements. See Szabadváry (1966) for more discussion.

combining mass ratio is a direct consequence of the fact that oxygen atoms are eight times heavier than hydrogen atoms.

Since the introduction of the atomic weight concept by Dalton, chemists are able to express their observations in a quantity that is proportional to the number of elementary entities. It became possible, for the first time, to establish a practical connection between mass measurements and chemical stoichiometry. The use of atomic weights, in conjunction with the mass measurements of substances, still remains a common means for stoichiometric calculations in chemistry (Meija 2014). A problem inherent to Dalton's approach, however, was the assumption of the chemical composition. For, if a water molecule contained *two* atoms of hydrogen and one oxygen atom, then the observed 8:1 combining mass ratio of oxygen and hydrogen gases corresponds, means that oxygen atoms are sixteen times heavier than hydrogen atoms. Thus, proper determination of atomic weights necessitated knowledge of chemical composition yet one could not determine chemical composition without proper atomic weights. This *catch-22* plagued chemistry for almost all of the 19th century despite the fact that 32 year old Italian chemist Stanislao Cannizzaro noted during the 1860 Karlsruhe Congress that the long-forgotten Avogadro hypothesis enables the breaking of this logical circle (Mönnich 2010).

Central to these developments in analytical chemistry was the discovery that matter reacting chemically does not do so simply between equal masses of the samples involved. We now refer to the study of this phenomenon by using Richter's term 'stoichiometry', now defined as the relationship between the amounts of substance that react together, and the products that are formed. Hence, chemists still carried out their measurements using an analytical balance, as they did before, but they now rationalized all combining masses through the corresponding relative atomic masses. The *Encyclopaedia Britannica* (1883, 9th edn, vol 16) explained this practice as follows:

> When a chemist speaks of acting on a molecule of succinic acid with two molecules of pentachloride of phosphorus, he means that he mixes them in the proportion of 118 parts of the former to 2 × 177.5 of the latter. For the sake of precision we sometimes speak of a molecule of water (or other substance) in grammes, or even of a *gramme-molecule*, a *grain-molecule, &c.* Thus, in the case just mentioned a gramme-molecule of succinic acid means 118 grammes of succinic acid, &c.

6.4 The many atomic weight scales

Indeed, throughout most of the 19th century chemistry was plagued with inaccurate atomic weights largely because of the differing opinions regarding the underlying constitution of the molecules. One also has to bear in mind the difficulty in accurate measurements of these values, which was eventually recognized with the 1914 Nobel Prize for chemistry to Theodore W Richards. There was, however, an additional source of discontent. While Dalton used hydrogen as the basis for the atomic weight

Table 6.1. Relative atomic mass scales through the centuries (Jensen and Meija 2010).

19th century: average mass scale	
H = 1 (O = 5.5)	Dalton, 1803–1805
H = 1 (O = 16)	Davy, 1812
O = 10	Wollaston, 1813
O = 1	Thomson, 1813, 1825
O = 100	Berzelius, 1814
O = 4	Griffin, 1834
O = 16	Brauner 1889; Clarke, 1893
C = 12	Hinrichs, 1893
20th century: isotopic mass scale	
$^{16}O = 16$	Aston, 1931
$^{12}C = 12$	IUPAC, 1961

scale, chemists were split on the merit of this choice for a long time. In fact, there have been numerous other scales used since Dalton, many of these alternative atomic weight scales used oxygen as the basis following a suggestion by Wollaston (see table 6.1).

By the end of the 19th century, two scales gained popular support: H = 1 and O = 16. This duality was undesired in science and one of the inaugural tasks of the newly-formed International Commission on Atomic Weights was to decide (in 1899) whether 'O = 16 shall be fixed as the future standard for the calculation of atomic weights'[4]. The Commission did side with oxygen but the debate was not over. In 1901, Theodore W Richards summarized the debate over the choice of hydrogen or oxygen as the atomic weight unit as follows (Richards 1901):

> One regrets that so much time should have been spent in discussing a matter which involves no fundamental principle, but is simply a question of form and of convenience.

In addition to educational features, the choice between hydrogen and oxygen as the atomic weight unit had significant practical implications. Czech chemist Bohuslav Brauner (1855–1935) showed that the measurements of the O:H ratio varied from 15.87 to 16.01 (Brauner 1889). Each new determination of this ratio gave a new value for oxygen because hydrogen was set to H = 1 by definition. Since most atomic weights were measured in relation to oxygen, and not hydrogen—owing to the trivial fact that oxides are more stable than hydrides—atomic weights of most elements could not be established to better than 0.5% uncertainty. To avoid this problem, Brauner suggested to return to the oxygen scale by setting O = 16.

[4] Not unlike the undergoing debates regarding the redefinition of the mole, one of the main oppositions to O = 16 scale was pedagogical. People claimed that it would be confusing to see the atomic weight of the lightest element, hydrogen, as 1.008, and not 1 exactly. For more discussion on this matter refer to (Richards 1900).

In 1920, during the first General Assembly of the International Union of Pure and Applied Chemistry (IUPAC) the question to reintroduce H = 1 was back on the agenda but was rejected by the International Atomic Weights Commission.

With the discovery of oxygen isotopes in the late 1920s, scientists realized that physicists have been, in fact, using the oxygen-16 as the mass standard, whereas chemists relied on the average atomic mass of all its isotopes. The difference was tiny (0.03%) but it was soon discovered that oxygen isotopic abundances were not constant in nature. As a result of the inherent uncertainty in the atomic weight scale, no chemical measurement could have been done to a precision better than a few parts in 10^5. With no obvious solution at hand, chemists continued using the O = 16 scale and physicists retained $^{16}O = 16$. This schism ended in 1961 when physicists and chemists both agreed to adopt a carbon-12 based scale for atomic masses through their respective International Unions[5].

6.5 The name: mole

Chemists did not have a proper name for the quantity that refers to the size of an ensemble of entities. As is often the case in science, developments in terminology and the symbolic language come second to technical advances. Only after the World War II the international interest in the symbols of physics became mature and scientists realized that 'gram-molecular weight' has the nature of a base quantity as it measures the size of the ensemble of atoms and molecules.

During the 1950s it became apparent that the mole was understood by chemists in two different meanings: as a certain *mass* of a substance (1 mol of potassium = 39 g) and as a certain *number of entities* (1 mol of potassium = 6.02×10^{23} atoms) (Milton and Mills 2009). Indeed, the word mole was introduced by two future Nobel laureates—Walther Nernst and Wilhelm Ostwald—as a practical way to compare the number of entities contained in a given bulk mass of a substance. The purpose of this mathematical construct has always been to compare the number of entities and this eventually became the official interpretation of the 'amount of substance' as was explained in the 1957 German Standard DIN 5484:

> Unter Stoffmenge wird im folgenden eine physikalische Grofienart verstanden, die dazu dient, einen aus bestimmten unter sich gleichen [...] Teilchen bestehenden Korper oder eine sonstige Gesamtheit solcher nach der Anzahl dieser Teilchen zu bewerten. [...] Demnach haben zwei Korper oder Gesamtheiten die gleiche Stoffmenge, wenn sie gleiche Anzahlen der jeweils gemeinten Teilchen enthalten.

> In the following text, the term 'quantity of substance' is understood to mean a physical quantity which is used to evaluate a body consisting of identical [...] particles or aggregates of particles according to the number of these particles.

[5] These developments are recounted by Henry E Duckworth in his memoir (Duckworth 2000).

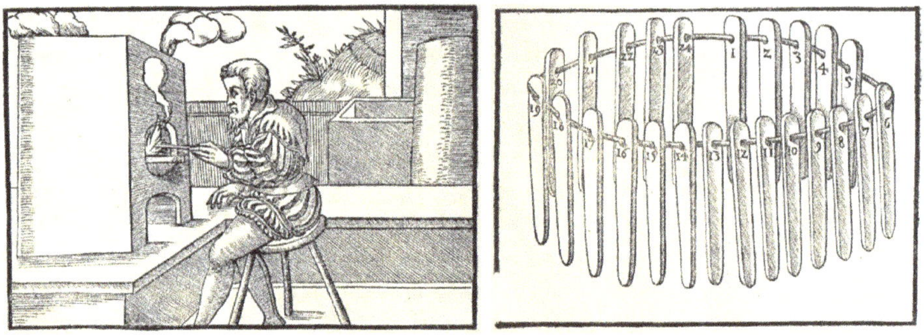

Figure 6.1. Two methods to determine the purity of precious metals which have been used since antiquity. (Left) The cupellation process, in which the object to be tested is melted down in a special manner and the mass of the purified residue is compared to the initial value. (Right) The purity of gold can be determined by comparing the colour of the material to that of a standard set of needles containing gold of known purity. Woodcuts from Georg Agricola's *De re metallica* (2nd edn, Basil, 1561).

> [...] Accordingly, two bodies or aggregates have the same amount of substance if they contain equal numbers of particles.

After more than a century in use, in 1971, the 'gram molecule' was to become a full-fledged unit with a proper name: the mole. This quantity was eventually called the 'amount of substance' and, in keeping with the long-held tradition of chemists, a quantity corresponding to 1 gram-molecule was chosen as the SI base unit with the name 'mole'. Following the advice of the International Union of Pure and Applied Physics, the International Union of Pure and Applied Chemistry, and of the International Organization for Standardization, the mole was officially adopted as a base unit of the SI by the 14th CGPM in October 1971[6].

The English name 'amount of substance' was derived from the German word *Stoffmenge* and it is not a well-liked name (Mills and Milton 2009). One problem with this quantity name is its unwieldy application by some, as in the 'amount of substance of benzene'. IUPAC notes that the word 'substance' in the 'amount of substance' is a placeholder for the actual substance which therefore renders 'amount of benzene'. This leads to another difficulty since the word 'amount' is too universal to be understood only in the chemical context. For this reason, many prefer a two-word 'chemical amount' which is similar to 'electric current' and also adds the word chemistry explicitly in the list of base quantities (Marquardt *et al* 2017). In fact, 'chemical amount' appears as an alternative name for 'amount of substance' in the IUPAC Green Book since 1993.

The word 'moles' is Latin for 'mass' and its modern diminutive 'molecula' has been used by many chemists to designate a particulate matter of 'little mass'. Thus, when Avogadro spoke of molecules in his seminal 1811 manuscript he meant the

[6] Three delegations (of nearly 40) cast votes against this Resolution: Poland, Czech Republic and USSR. See, for example, the article by Aleksandrov Yu I for further discussion about the mole.

smallest particles. Much like the Greek philosophers of antiquity, we use the word 'atom' to describe the same concept.

With the advent of the atomic theory during the 19th century, chemists were progressively distinguishing and describing matter at two levels: molar and molecular. August Wilhelm von Hofmann made use of the word 'moles' and in his 1865 textbook *An Introduction to Modern Chemistry* he introduced the adjectival form 'molar' which since has become synonymous with chemistry:

> the reciprocal actions of minute particles through insensible intervals of space are distinguished as *molecular*. We may fairly therefore contradistinguish, by the epitet *molar*, the reciprocal actions of measurable masses through measurable intervals of space. [p 140; emphases in the original]

The adjectival form 'molar' has become commonplace in chemistry and is joined with decimal prefixes as in 'decimolar' or 'millimolar'. In the mid-nineteenth century it was common to use the phrase 'gram-molecular weight' in chemistry to denote the mass of a substance that is equal to its molecular weight. As time went by, this unwieldy phrase was variously shortened to 'gr.mol.wt, gr.mol, or g-mol. which eventually became abbreviated to 'mol.' or 'Mol.'. In the late 1890s, Ostwald and Nernst's proposal was to change this abbreviation into a stand-alone word 'Mol' while retaining the same historical meaning of gram-molecule. Ostwald wrote in 1893 (Ostwald 1893):

> Nennen wir allgemein das Gewicht in Grammen, welches dem Molekulargewicht eines gegebenen Stoffes numerisch gleich ist, ein Mol [...]

> Let us generally refer to the weight in grams that is numerically identical to the molecular weight of that substance, as one mole [...]

In addition to 'Mol', Nernst also used the word 'Mole'. Although mole can be seen as a logical simplification of the gram-molecule, the acceptance of this term was not swift. The early English translations of Ostwald's and Nernst's textbooks omitted this terminological proposal and generally reverted 'Mol' back to 'g.-mol.' (Ostwald 1894, Nernst 1895). It was not until Alexander Findley's translation of Ostwald's *Inorganic Chemistry* in 1902 when 'mole' first appeared in English texts. IUPAC recommended 'mol' as the symbol for mole in 1963 (Comptes Rendus XXII Conference 1963).

6.6 Molar measurements in practice

In the early 19th century, the results of chemical analyses and calculations were done on the mass basis. Hence, different scales were required to compare, say, the amount of soda or potash. A prominent science writer of his time, Scottish chemist Andrew Ure (1778–1857), describes, for example, a 'normal solution of sea salt' as a solution

Figure 6.2. (from left) Dalton, Avogadro, Berzelius, and Cannizzaro: four key figures in the advancement of chemical measurements in 19th century. Although Dalton was not the first to propose the atomic theory, he put forward that atoms come in different weights. Dalton also formulated the law of multiple proportions which could be easily explained with the help of the atomic theory. Avogadro noted that equal volumes of gases must contain equal number of atoms. Hence, elucidation of stoichiometry and atomic weights could be greatly facilitated by measurements of combining volumes of substances and not just their masses. Berzelius is especially noted for the development of classical analytical techniques and precise determinations of atomic weights. 'Berzelius gave order to everything he touched', as Ronald G W Norrish noted in his Nobel Prize acceptance speech. With Cannizzaro's advance, chemists finally acquired a standard set of atomic weights and were able to determine unambiguous and universally accepted compositional formulas for their compounds. Caricatures courtesy of William B Jensen (University of Cincinnati).

'of which 100 grammes will precipitate exactly one gramme of silver' (Ure 1848). Such an approach to chemical analysis generally required that each substance to be tested had its own reagent prepared in a concentration that would provide conveniently the analysis results as the mass fraction (percent).

The English chemist John Joseph Griffin noted another shortcoming to many 19th century chemical measurements: the common use of density of a chemical solution to ascertain its strength is an unacceptable proxy (Griffin 1848, p 103, Griffin 1851). He noted that for many common chemicals the density of their solutions is simply a poor and ambiguous substitute for 'chemical strength'

> All who have mastered the elements of theoretical chemistry know, that the *power* of chemical solution depends upon the *number* of atoms or equivalents it contains, and not upon the *absolute weight* of those atoms. Diluted nitric acid of 300° is twice as strong as diluted sulphuric acid of 150°, because there are twice as many chemical atoms present in it, not because the atoms of the nitric acid weigh twice as much as those of the sulphuric acid.

Due to the peculiar relationship between the concentration and density of its solutions, acetic acid of density 1.065 to 1.066 g mL^{-1}, for example, can be either 60% or 90% strength. In addition, 1% difference in the densities of ammonia solutions (0.95 vs 0.96 g mL^{-1}) corresponds to nearly 30% difference in the mass fraction of ammonia. Griffin advocated the use of 'chemical strength' to achieve greater uniformity and clarity. His 'centigrade testing' relied on preparation of

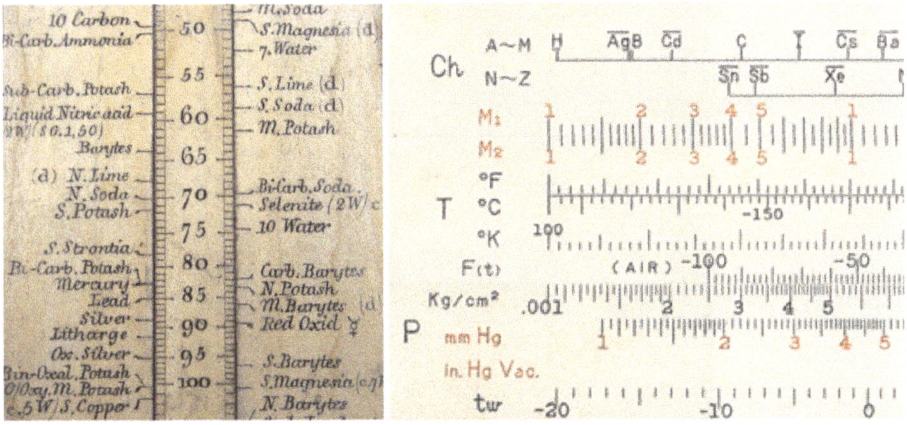

Figure 6.3. Wollaston's 1814 slide rule for calculating molecular weights (left). Only six original slide rules are believed to exist today. Similar slide rules were still being manufactured during the 1950s such as the Sun Hemmi 257—For Chemical Engineer (right). In Hemmi 257, placing the cursor over an element or chemical group on the upper scale allows the molecular weight to be read on the scale located on the on other side of the slide rule. Courtesy of Tesseract—Early Scientific Instruments.

chemical solutions of 'equivalent strength' (Griffin used the centigrade notation as in '100° strong'). Such solutions were prepared by dissolving 'one test atom of the chemical preparation in so much water as will make a decigallon of solution at 62° Fahr'. The success of this approach was in its inherent feature that equally strong acids and bases will neutralize one another in equal volumes. The use of atomic weights (chemical equivalents) in reporting chemical results was further popularized by Karl Friedrich Mohr in his seminal 1855 textbook which formed the basis for volumetric chemical analysis (Mohr 1855).

The use of the term 'molarity' was preceded in time and popularity by 'normality'. In fact, since the mid-19th century chemists frequently applied the phrase 'normal solution' to denote solutions that 'contain one atomic weight of the active chemical, weighed in Grammes, and dissolved in a Litre of solution'. Later, the normality would also be adjusted for the stoichiometry of the chemical reaction for which such solution is intended (Haynes 1895). Chemists have been, in fact, using the term 'normal' to describe solutions containing one 'equivalent' of substance in a litre of solution.

Both systems had their advantages and disadvantages. In the old, mass-based, system, a statement '7.1 mg of phosphorus oxide' is timeless and unambiguous. In the new, amount-based, system, however, the results of chemical measurements were no longer based solely on the actual masses of reactants. Instead, the comparison of the various substances was achieved from their mass via atomic weights. This becomes problematic if unreliable atomic weights are employed. Indeed, the atomic weights themselves did not become reliable until the late 19th century. Fleischer's *A System of Volumetric Analysis* (Fleischer 1877) illustrates the unwanted consequences of a theory-laden with reliance on atomic weights:

Figure 6.4. Grapefruit-sized sphere made from a single crystal of nearly pure silicon-28. The sphere weighs 1 kilogram and was used to determine the Avogadro constant (and the Avogadro number) with an unmatched accuracy. Photo credit: Olaf Rienitz, PTB.

> Thus, 1 cb.c. [mL] of normal hydrochloric acid containing 36.5 m.gm [mg] HCl neutralizes an equivalent of caustic soda in m.gms.—i.e., 40 m.gm.; 2 cb. c. neutralize 80 m.gm. NaHO, and so on. (In old notation 1 cb.c. normal acid neutralizes 32 m.gm. NaHO.)

If the amount of HCl is used as a measure of NaOH, we witness here a 20% error based on the change of the atomic weight of oxygen from O = 8 to O = 16 which subsequently changed the relative molecular weight of NaOH from 32 to 40 (in the scale H = 1). Problems with amount-based notation did not stop with atomic weights. The use of 'normality' has been deprecated for many decades largely due to its core ambiguity: 1 N $BaCl_2$ solution could refer to a 1 N barium(II) solution or 1 N chloride solution. The former corresponds to 0.05 M $BaCl_2$, whereas the latter equates to 1 M $BaCl_2$.

Although Dalton put forward the atomic theory, his theoretical views were not readily met with acceptance. Interestingly enough, Dalton's atomic theory owes much of its success to a wooden slide rule—a mechanical calculator of its time—which was introduced by English chemist William Hyde Wollaston in 1814 (Wollaston 1814). Wollaston's slide rule contained the chemical equivalents (relative combining weights of elements) on the base of the rule and on the sliding part of it. Because both scales were logarithmically spaced, the slide rule allowed for a quick calculation of the relative masses of substances reacting with one another, the quantity of products, or the relative proportion of elements in a compound (Williams 1992). Wollaston's slide rule soon became an indispensable item of laboratory equipment which has prompted some chemists even to proclaim that it has facilitated the development of chemical analysis more than any other invention (Comstock 1834).

6.7 Amount of substance as a dimensional quantity

Discussions on 'dimensions' are quite controversial and often subjective (Emerson 2005). Edward A Guggenheim noted in 1942 that 'for special problems it may be advantageous to increase the number of fundamental quantities above the usual number. It can sometimes be useful in dimensional analysis to regard the number of atoms as having dimensions different from a pure number' (Guggenheim 1942). Thus, conferring a unique 'dimensionality' to the amount of substance marks parallels to temperature which we do distinguish from thermal energy as a matter of convenience.

Nevertheless, the mole is often said to be an arbitrary unit. Such comments are unhelpful because all units are arbitrary. The kilogram did not have to be tied to the mass of one litre of water, the metre did not have to be tied to a quadrant of Earth, and the kelvin did not have to be tied to the triple point of water. Likewise, the mole did not have to be tied to the mass of carbon-12. All decisions that have set the magnitude of base units are results of practical, albeit arbitrary, decisions.

6.8 The Avogadro number

Scientists had been wondering about the size of atoms and molecules for a long time. Since gram-molecule had been established as a natural unit of mass when dealing with chemical substances, it was natural to wonder about 'the number of actual molecules contained in one gram-molecule', as Einstein did in 1905. In 1909, the future Nobel laureate Jean Perrin proposed calling this number in honour of Amedeo Avogadro (Perrin 1909):

> Ce nombre invariable N est une constant universelle qu'il semble juste d'appeler *constant d'Avogadro.*

> This invariable number N is a universal constant and it seems fair to name it *Avogadro's constant.*

While today we distinguish carefully between the Avogadro constant and the Avogadro number, this terminological and conceptual distinction is rather new. Many have argued that this number (or constant) is not a fully-fledged physical constant; rather, a man-made arbitrary scaling factor or a 'constant of a lesser breed' (Mills 2010). On the contrary, some have called it 'the most important of all physical constants' (Hinshelwood 1956) During the 1926 Nobel Prize Award Ceremony Speech, Professor Carl Wilhelm Oseen, member of the Nobel Committee for Physics[7], had this to say: *Perrin was able to determine one of the most important physical constants, Avogadro's number.* Whether we like it or not, Avogadro number is an important aspect of modern science. One of Einstein's three *annus mirabilis*

[7] Among many other nominators, Oseen successfully nominated Albert Einstein for the 1921 Nobel Prize for Physics.

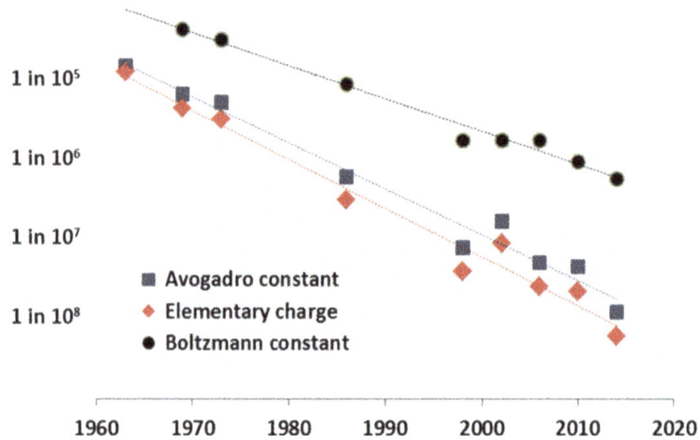

Figure 6.5. Advances in science manifest as ten-fold reduction in the uncertainty of most fundamental physical constants in every two decades or so, as shown here in the case of the Avogadro constant (N_A), the elementary charge (e), and the Boltzmann constant (k_B). Today, fundamental physical constants are now known with sufficient certainty to redefine the International System of Units in terms of some of these constants.

papers dealt with Brownian motion and ways to best determine the Avogadro number.

The Avogadro number remains relevant in science because it is a scaling factor between two mass units still used in science: the kilogram and the dalton, the latter being the 1/12 mass of a single carbon-12 atom. Determination of the Avogadro constant plays an important role in science because it enables a comparison of disparate experiments of other constants. For example, the Rydberg constant relates the Avogadro and Planck constants through several other well-known physical constants. Because of this, the value of the molar Planck constant ($N_A h$) is known better than either N_A or h and therefore determination of the Avogadro constant indirectly provides a value of the Planck constant (Becker and Bettin 2011). In the early 1990s, several of the worlds' leading metrology institutes started the work on determining the value of the Avogadro number using x-ray crystal density method. Here, the density of a material is measured in two ways, at the macroscopic and atomic levels:

$$\rho_1 = m_{\text{sphere}}/V_{\text{sphere}}\ [\text{kg/m}^{-3}]$$
$$\rho_2 = m_{\text{unit cell}}/V_{\text{unit cell}}\ [\text{Da/m}^{-3}]$$

Equating these two density measurements provides the value of the Avogadro number, $\{N_A\}$ = kg/Da, which is the numerical value of the Avogadro constant. This expensive experiment remains the most accurate realization of the definition of the mole to date.

6.9 Proposed new definition of the mole

Chemists rely on mass measurements and relative atomic masses (atomic weights) of atoms to enable them to conceptually realize the amount of substance. This is why chemical measurements are almost invariably traceable to the relative atomic masses.

In the early 20th century the Avogadro number was viewed as a scaling factor for mass. When referring to the Avogadro number, for example, Percy Williams Bridgman notes that 'its dimensions are evidently the reciprocal of a mass' (Bridgeman 1922). Today, we view its dimension as the reciprocal of the amount of substance. This paradigm shift aligns with the common view among chemists that 'the mole is the Avogadro number of entities' as it was summarized by the IUPAC ICTNS in 2009. It is not easy to gauge public opinion on a broad technical matter such as the redefinition of the mole. However, in the last two decades, significant support has been generated for a definition of the mole based on a fixed number of entities (Mills 2006, Milton and Mills 2009).

> The mole is the amount of substance of a system that contains exactly $6.022\ 1415 \times 10^{23}$ specified elementary entities, which may be atoms, molecules, ions, electrons, other particles or specified groups of such particles.

(The precise value for the Avogadro number to be used in the definition of the mole will be set by the CODATA Task Group on Fundamental Constants prior to the 26th CGPM in 2018.) At core, the redefinition of the mole centred on the question of whether it should be defined as an amount of substance contained in a certain *mass* of unbound carbon-12 atoms or a certain *number* of entities. In both cases, the magnitude of the mole remains unchanged.

6.10 Consequences of the entity-based definition

Chemists enjoy the relationship $1\ \text{g mol}^{-1} = 1\ \text{Da/ent}$ which states that the atomic mass of entities (atomic weight) is numerically identical to the mass of one mole of such entities when it is expressed in grams. This relationship is true because the current definition of the mole has the effect of setting the molar mass of carbon-12 to $0.012\ \text{kg mol}^{-1}$ and because the current definition of the dalton is based on the mass of the carbon-12 atom, $m_a(^{12}\text{C}) = 12$ Da. Once the mole is no longer tied to carbon-12, the exact link between the atomic weights and molar masses will no longer be there. In other words, 12 g of carbon-12 will no longer be exactly 1 mol (of carbon-12).

The molar mass of a substance is related to its molecular weight via $M(\text{X}) = A_r(\text{X})\ M_u$ where M_u is the molar mass constant with an exact value of $10^{-3}\ \text{kg mol}^{-1}$. In the new SI, M_u will no longer have an exact value but rather will be set via the dalton as $M_u = N_A m_u$. CODATA-2014 puts the uncertainty of M_u to 1.2 parts in 10^{10}. A discrepancy in the molar mass at the level of one part in 10^{10} is not in the realm of concern for chemists. In fact, there are only two elements whose standard atomic weights are currently known with precision below a few parts in 10^{10}: fluorine and phosphorus. Hence, the fact that the molar mass of carbon-12 will now have an uncertainty of one part in 10^{10} will have a marginal impact on the molar masses of only a handful of substances such as F_2, P_4, or PF_3.

The above comments notwithstanding, many are surprised to find out that there is actually no *exact* link between molar masses and atomic weights. In other words, 0.012 kg of pure carbon-12 is, in fact, not exactly 1 mol under the 1971 definition of the mole if the carbon is in solid form, and at room temperature. This is because in the definition of the mole, it is understood that unbound atoms of carbon-12, at rest and in their ground state, are referred to. Chemists do not work with unbound atoms; and atoms are at rest and in their ground state only at zero kelvin temperature, whereas chemists normally perform their measurements at room temperature. In fact, the molar mass of crystalline substances is given by

$$M(\mathrm{X}) = A_{\mathrm{r}}(\mathrm{X})M_{\mathrm{u}} - \Delta_{\mathrm{f}}H_{\mathrm{c}}^{0}(\mathrm{X})/c_0^2$$

where $A_{\mathrm{r}}(\mathrm{X})$ is the molecular weight (sum of all relevant atomic weights), M_{u} is the molar mass constant, $\Delta_{\mathrm{f}}H_{\mathrm{c}}^{0}(\mathrm{X})$ is the cohesive energy of the crystal, and c_0 is the speed of light in vacuum. For a graphite crystal, $\Delta_{\mathrm{f}}H_{\mathrm{c}}^{0}(\mathrm{X})$ = 711 kJ/mol which corresponds to a difference between the molar mass of bound and unbound graphite of almost one part in 10^9 (equivalent to 1 μg in 1 kg). Currently, chemists mostly ignore the discrepancy between 1 g/mol and 1 Da at the level of one part in 10^9 so there is no reason to believe that a ten-fold smaller discrepancy will become problematic. Thus, while 1 mol of substance always contains the same number of specified entities, the 1971 definition requires corrections to be made for bonding energy, whereas the entity-based definition does not (Davis and Milton 2014). In this sense, the definition of the mole in the new SI is more fundamental, than the 1971 definition.

Atomic weights and chemical calculations have become so intertwined with chemical measurements that it becomes natural to wonder if there is a way to realize the new definition of the mole without invoking the atomic weights. In 2010, for example, Schlegel and coworkers described an experiment where bismuth ions were accumulated on a metal disk (Schlegel *et al* 2010). A total of 323.1 mg of bismuth was accumulated over 24 h requiring a total electric charge of 149.2 coulomb (ampere seconds). Since one mole of electrons corresponds to a charge of $F = N_{\mathrm{A}}e$ = 96 485 A s, the 323.1 mg of bismuth therefore corresponds to 1.55 mmol of bismuth. In this experiment, the mole is effectively realized from definitions of the ampere and second. Alternatively, we find that for bismuth the ratio of its mass and chemical amount is 323.1 mg/1.55 mmol = 209.0 g mol^{-1} from this experiment.

6.11 Outlook

The mole is thought of by many chemists as a quantity that contains the Avogadro number of entities. The upcoming revision of the International System of Units will align the official definition of the mole with this commonly held view. While the new definition does offer technical improvements over the 1971 definition, virtually no chemist will be able to take advantage of it in the foreseeable future. To them, the biggest advantage will certainly be in the formulation of the mole which will specify the exact number of entities. The techniques that can be used for realization and dissemination of the mole will remain the same as before. These include gravimetry (with corrections for chemical purity), electrolysis, and use of the ideal gas law (with corrections for non-ideality). Under best practices, these three methods can achieve precision from few parts in 10^4 to few parts in 10^6 for realizing the amount of substance (although some state-of-the-art measurements can do better, as described in the above x-ray crystal density method for silicon). In the new SI, Faraday constant and the universal gas constant will acquire fixed numerical values, whereas the molar mass of carbon-12 will no longer be exact. However, the uncertainty on this value will be less than one part in 10^9 with no consequence to chemists, thereby allowing continuity with the previous definition.

References

Becker P and Bettin H 2011 *Philos. Trans. A Math. Phys. Eng. Sci.* **369** 3925–35

Boyle R 1725 *The Philosophical Works of the Honourable Robert Boyle. Vol. II, ed Peter Shaw (London 1725)* Experiments and observations upon colours (W and J Innys)

Brauner B 1889 Die Basis der Atomgewichte *Ber. Dtsch. Chem. Ges.* **22** 1186–92 See also Chem. News 1888, 58, 307–8

Bridgman P W 1922 Dimensional analysis (New Haven, CT: Yale University Press)

Comptes Rendus XXII Conference London 1963 (London: Butterworths Scientific)

Comstock J L 1834 Elements of Chemistry, reprint 2016 (Wentworth Press)

Dalton J 1808 *A New System of Chemical Philosophy*

Davis R S and Milton M J T 2014 *Metrologia* **51** 169–73

Duckworth H E 2000 *One Version of the Facts: My Life in the Ivory Tower* (Winnipeg: University of Manitoba Press)

Emerson W H 2005 On the concept of dimension *Metrologia* **42** L01

Fleischer E 1877 *A System of Volumetric Analysis*reprint 2016 (Wentworth Press)

Frankland E 1878 Extract from a Letter of Dr Frankland to Mr Sylvester *Am. J. Math.* **1** 345–9

Griffin J 1848 Chemical Reactions: a Popular Manual of Experimental Chemistry (London: John Joseph Griffin)

Griffin J 1851 On the use of Centigrade testing in pharmacy *Pharmaceut. J.* Jan **1851**

Guggenheim E A 1942 Units and Dimensions *Philos. Mag.* **33** 479–96

Hartog P J 1894 *Nature* **50** 149–50

Haynes D O 1895 *The Phamaceutical Era* vol 14 (D O Haynes and Co.)

Hinshelwood C N 1956 Amedeo Avogadro *Science* **124** 708–13

Homberg 1699 Observation sur le quantité exacte des Sels Volatils Acides contenus dans les differens Esprits Acides *Histoire de l'Academie Royale des Sciences* **1** 44

Jensen W B and Meija J 2010 *Anal Bioanal. Chem.* **398** 11–2

Kuhn T S 1961 The Function of Measurement in Modern Physical Science *Isis* **52** 161–93

Lavoisier 1790 *Elements of Chemistry*, Engl. transl. R.Kerr

Marquard R *et al* 2017 A critical review of the proposed definitions of fundamental chemical quantities and their impact on chemical communities (IUPAC Technical Report) *Pure Appl. Chem.* https://doi.org/10.1515/pac-2016-0808

McNulty M B 2014 Kant's Philosophy of Chemistry *Doctor of Philosophy dissertation* (Irvine: University of California)

Meija J 2014 An ode the atomic weights *Nat. Chem.* **6** 749–50

Mills I and Milton M 2009 Amount of substance and the mole *Chem. Int.* **31** 3–7

Mills I M 2010 What is a mole?: old concepts and new *Chem. Int.* **32** 1

Mills I M *et al* 2006 Redefinition of the kilogram, ampere, kelvin and mole: a proposed approach to implementing CIPM recommendation 1 (CI-2005) *Metrologia* **43** 227
Milton M J T and Mills I M 2009 Amount of substance and the proposed redefinition of the mole *Metrologia* **46** 332–8
Mohr F 1855 Lehrbuch der chemisch-analytis chen Titrirmethode (Berlin: F Vieweg)
Mönnich M W 2010 Thriving for unity in chemistry: the first international gathering of chemists *Chem. Inter.* **32** 10–4
Nernst W 1895 *Theoretical Chemistry* (transl.) ed C S Palmer (London: Macmillan)
Ostwald W 1893 *Hand- und Hilfsbuch zur Ausführung Physiko-Chemischer Messungen* (Leipzig: Engelmann) p 119
Ostwald W 1894 *Manual of physico-chemical measurements transl.* ed J Walker (London: Macmillan)
Perrin J 1909 Mouvement brownien et réalité moléculaire *Ann. Chim. Physique* **18** 5–114
Pliny the Elder The Natural History Book XXXIII ch 43 ed J Bostock
Prout W 1815 On the relation between the specific gravities of bodies in their gaseous state and the weights of their atoms *Ann. Philos.* **6** 321–30
Rainey A F Z L 2015 Amarna Letter EA7. The El-Amarna Correspondence vol 1 (Leiden, Boston: Brill)
Richards T W 1900 International atomic weights *Proc. Am. Acad. Arts Sci.* **36** 171–6
Richards T W 1901 The standard of atomic weights *Proc. Am. Acad. Arts Sci.* **37** 177–81
Schlegel C *et al* 2010 The determination of the atomic mass constant with ion accumulation: status and perspectives *Metrologia* **47** 146
Smartt Bel M 2005 Lavoisier in the Year One: The Birth of a New Science in an Age of Revolution (New York: Norton)
Szabadváry F 1966 History of Analytical Chemistry: International Series of Monographs in Analytical Chemistry vol 26 (Oxford: Pergamon)
Ure A 1848 A Dictionary of Arts, Manufactures, and Mines (New York: D Appleton and Co.)
Wälchli W 1981 Touching Precious Metals *Gold Bull.* **14** 154–9
Williams W D 1992 Some early chemical slide rules *Bull. Hist. Chem.* **12** 24–9
Wollaston W H 1814 A Synoptic Scale of Chemical Equivalents *Philos. Trans. R. Soc.* **104** 1–22

IOP Publishing

Precise Dimensions
A history of units from 1791-2018
Malcolm Cooper and Jim Grozier

Chapter 7

The history of the SI unit of light, the candela

7.1 Introduction: light and vision

Of our five senses, vision is arguably the most useful and important. It is chiefly through sight that we understand our environment and gain the information we need to govern our actions and movements. Furthermore, our ability to carry out most tasks in a safe, effective and efficient manner is dependent to a large extent on the ease with which they can be seen. This reliance on vision is amply demonstrated by the fact that about 30% of the human cerebral cortex is dedicated to visual analysis and perception, as compared with around 8% for touch and just 3% for hearing.

The ability to produce and control light has developed dramatically over the last 100 years or so, and during this period the provision of lighting has come to be regarded as a basic necessity of life. In the developed world we take it for granted not only that light is available at the flick of a switch, but also that the quality of the light provided will be suitable to allow us to perform the task at hand, whether it be operating machinery, driving a car, or simply reading a book. However, light not only stimulates vision, but can affect our mood, change our perceptions and directly influence our behaviour and emotions in many complex and unexpected ways. For example, most pedestrians report a greater feeling of safety when walking on a well-lit street at night as opposed to one which is not lit, and shops, restaurants and the like invest heavily in designing a lit environment that creates the right ambience, since they know that this impacts directly on consumer choices. Light can even control our health and well-being, playing a key role in areas ranging from sleep disorders through to the initial onset and/or subsequent growth rates of various cancers.

Given the critical role that vision plays in our daily lives, it is obvious why such great importance should be attached to providing adequate and effective lighting, and to the ability to measure a lit environment to confirm it achieves the required performance. It is also not surprising that there should be a desire to quantify light in a way that relates to its visual effect (i.e. in units that include consideration of human

doi:10.1088/978-0-7503-1487-9ch7 © IOP Publishing Ltd 2017

physiology) rather than in purely physical terms. This chapter will explore the history of the measurement of 'light' and examine how this led to the evolution of units within the International System of Units (SI) relating specifically to visual effectiveness.

7.2 Artefact-based standards and units for measurement of 'light'

7.2.1 The evolution of artificial lighting

The history of the measurement of light is inextricably bound up with the historical development of artificial sources of light. This began with flame sources, initially in the form of candles and oil lamps and later supplemented by coal gas lighting systems. However flame sources are not only inefficient, since much of the available energy either escapes before it can be converted into radiant energy or is wasted in the form of heat rather than visible radiation, but they are also a potential fire risk. The first major advance in 'safe' artificial lighting came with the invention of the electric carbon arc in the mid-1800s, in which light was produced by incandescence at a high temperature, providing an intense, efficient, white light source. Indeed this was so effective that it was then considered too powerful for general use. This was soon superseded by tungsten filament incandescent lamps, which can not only operate at a high temperature due to the high melting point of tungsten, but also have a higher ratio of visible to infrared radiation than a carbon filament (or other non-selective radiator at the same temperature) due to the particular emissive properties of tungsten. Tungsten filament lamps are compact, convenient and relatively cheap to produce and operate, and as such were the mainstay of lighting in homes in particular until very recently. They have the further advantage that they provide radiation at all wavelengths across the whole visible spectrum and thus have excellent colour rendition properties. However, in applications where colour discrimination is not critical and energy efficiency is a more important consideration, other light sources, such as sodium and fluorescent lamps, soon gained supremacy. In these devices electrical energy is used to excite electrons in a gas vapour, which then decay back to the ground state either directly emitting visible light in the process or, in the case of fluorescent lamps, emitting radiation in the UV region to stimulate a phosphor which then emits radiation in the visible region. Although these are highly efficient compared with incandescent lamps, since energy is not wasted in the form of heat, they have rather poor colour quality since the output radiation is confined to the lines of the emission spectrum of the element in question (sodium, for example) or the spectral emission profile of the phosphor(s) used. The most recent major development in lighting is the LED, in which light is produced by electroluminescence in a semiconductor p–n junction, the spectral properties of the output radiation being determined by the semiconducting materials used. This leads to a light source that is compact, robust and energy efficient, with a long lifetime and (through combining several different colours of LED in a single device, often coupled with the use of phosphors) good colour rendition properties. Having started in niche areas such as numeric display panels and indicator lights, LEDs are now rapidly displacing all other light sources across most applications, from general purpose lighting in homes, offices, schools, shops and factories, to more specialist applications such as traffic signals and street lights.

7.2.2 Flame standards

A standard source should, if possible, be more stable than the sources to be measured against it. In order to be useful and acceptable as the basis for an agreed unit of light, it must also be able to be closely reproduced from one occasion to another and one location to another, i.e., it must be reproducible from a specification. Since the earliest light sources were flame sources, these were naturally also the earliest standards. Candles were used initially, made from high purity wax (e.g. a pure spermaceti candle weighing one sixth of a pound and burning at a rate of 120 grains per hour was used in England), but these were soon superseded by specially designed flame standards, burning oil at a measured rate or burning a defined mixture of a flammable gas and air. The most successful of these was the pentane lamp, which was invented by Harcourt in 1877 [1] and had an intensity roughly equivalent to that of one candle; this was later redesigned in a larger form with an intensity of about ten candles [2] (figure 7.1). This burned a mixture of pentane gas and air, with the rate of flow of the mixed vapour, and therefore the height of the flame, being controlled by the means of two stop-cocks. Although the intensity of the pentane lamp was found to depend on atmospheric pressure and humidity, the effect of these could be allowed for using a formula, so giving a standard that could be reliably reproduced from a very exact specification. From about 1900 on this was used regularly as a standard by the gas and other industries, as well as for much of the early photometric (i.e. light measurement) work at the UK's National Physical Laboratory (NPL). The only other flame standard of similar importance was the Hefner lamp [3], which burned pure amyl acetate with a flame height set accurately using a simple optical projection lamp; this was used as the official standard in Germany and some other European countries from the 1890s until the 1940s.

Flame standards suffered from problems of reproducibility and short term variability, so the advent of incandescent lamps, with their much more stable light output, naturally led to them being considered as possible standards. However, it was found impossible to manufacture incandescent lamps sufficiently reproducibly to enable their use as a primary standard and although research continued for several decades towards the end of the 19th century, only one incandescent source was adopted as a reference artefact. This was based on the intensity of the light produced by one square centimetre of a surface of platinum at the temperature of solidification, which was found to be reproducible to within one percent under certain conditions [4–6]. This so-called Violle standard (named after the J Violle who constructed the first practical version) was adopted by the International Electrical Congress in 1889, and in 1919 one twentieth part of this intensity (termed the 'bougie decimale') was adopted as the legal unit of luminous intensity in France.

Thus in the early part of the 20th century the situation in relation to standards for the measurement of light was highly complex: different countries used different

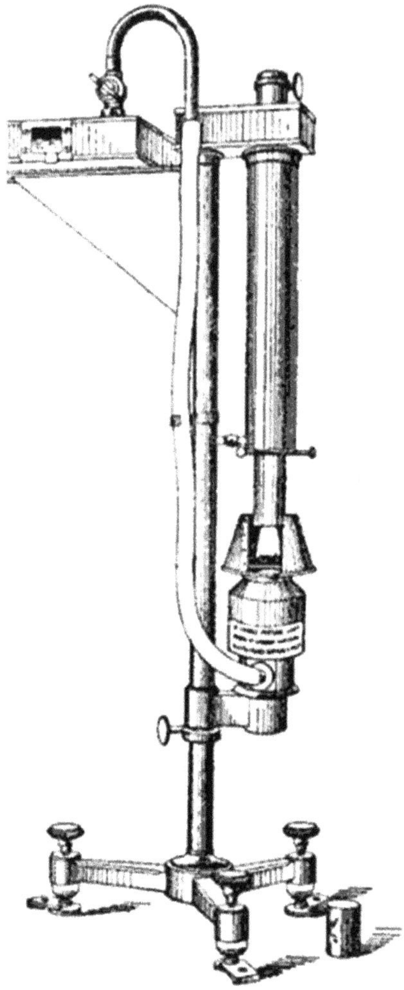

Figure 7.1. Harcourt pentane-air lamp. This drawing of Vernon Harcourt's air-pentane lamp has been obtained by the author(s) from the Wikimedia website Wikipedia https://commons.wikimedia.org/wiki/File:Harcourt_pentane_air-lamp.jpg, where it is stated to have been released into the public domain. It is included within this article on that basis.

standards, the legally-defined standards (often based on flame sources) were difficult to use and therefore not widely applied, and newer light sources, although easier to use, had not been adopted as legal references. To make matters even more complicated, it was found that the various standards in use differed quite appreciably from one another, making measurements completely incompatible. In 1909, therefore, an international agreement was signed between the National Bureau of Standards (USA), Laboratoire Central d'Électricité (France) and the National Physical Laboratory (UK) to adopt a common one-candle unit [7]. Although derived from flame standards, chiefly the ten-candle pentane lamp, the agreed value of the unit was maintained by groups of carbon filament lamps held at each of the three laboratories. Germany was not a signatory to this agreement and continued to use the 'Hefnerkerze' (based on the Hefner lamp) until 1942; the value of this was stated in the agreement to be nine-tenths of the agreed unit based on the results of comparison measurements. In 1921 the agreement was extended to include Belgium, Italy, Spain and Switzerland and the unit—still maintained in groups of filament lamps—was named the 'international candle [8].

7.2.3 The black-body standard, the 'new candle' and the 'candela'

Although the adoption of the international candle allowed countries to make measurements on a common basis, the reliance on groups of lamps, which inevitably deteriorated slightly each time they were used, cast grave doubt on the ability to guarantee the long-term consistency of the measurement scale. There was a keen desire to introduce instead a standard that could be reliably established on a repeatable basis, with no possibility of long-term drift in its value. Thus in 1930 the Comité International des Poids et Mesures instigated an investigation into using the intensity of a specified area of a black-body radiator at the melting point of platinum as such a reference, i.e., a source similar to the Violle standard. This involved the National Bureau of Standards, National Physical Laboratory and the University of Strasbourg each setting up their own standard of this form and comparing it with their existing filament standards. The values assigned to the platinum black-body standard by each laboratory, respectively, were 58.86, 59.00 and 58.78 international candles per square centimetre [9–11], leading the CIPM to adopt the following resolution at the inaugural meeting of the new Comité Consultatif de Photométrie (which later became the Comité Consultatif de Photométrie et Radiométrie, CCPR) in 1937 [12]:

> From January 1, 1940, the unit of luminous intensity shall be such that the brightness of a black-body radiator at the temperature of solidification of platinum is 60 units of intensity per square centimetre. This unit shall be called the 'new candle' (with appropriate translation into other languages).

This new definition was adopted not only by the signatories to the previous 'international candle' agreement, but also The Netherlands, Germany, Japan and the USSR. For Germany this was a major change, meaning not only abandoning the Hefnerkerze but also accepting a unit about 10% different in size. Immediately after the resolution was adopted, NPL was given the job by CIPM of carrying out a full comparison of standard lamps from five national laboratories, involving two groups of lamps from each laboratory operating at two different temperatures. The ten values of the new candle obtained in the course of this comparison ranged from 0.993 to 1.008 times the overall average [13], clearly demonstrating a very satisfactory level of international agreement.

Unfortunately, international events intervened and it was not until 1946 that the new candle was finally officially adopted by the CIPM [14]. Early in 1948 the

International Commission on Illumination (CIE) adopted the Latin name 'candela' (abbreviation cd) [15] for the new candle and this was later confirmed by the CIPM. The final definition adopted by the 9th General Conference on Weights and Measures (CGPM), 1948, was:

> Candela. The unit of luminous intensity. The magnitude of the candela is such that the luminance of a full radiator at the temperature of solidification of platinum is 60 candelas per square centimetre.

Subsequently the 10th CGPM, in 1954, approved the candela as a base unit within the Metre Convention due to the importance and significance of photometry within a practical system of units of measurement, and it continues today as one of the seven base units of the SI. The definition was amended slightly at the 13th CGPM in 1967, although this was simply a clarification and reformulation and did not change either the fundamental meaning of the definition, the size of the unit, or the techniques used for its realisation. The revised definition [16] was:

> The candela is the luminous intensity, in the perpendicular direction, of a surface of 1/600 000 square metre of a black body at the temperature of freezing platinum under a pressure of 101 325 newtons per square metre.

7.3 A radiometric approach to photometry

Although the introduction of the black-body definition of the candela represented a major step forward in the measurement of light, providing a reliable and reproducible basis for measurements, it suffered from a number of significant drawbacks. Firstly, it was found that the black-body standard was difficult and inconvenient to use (showing large variations in luminance from one freezing of the platinum ingot to the next, significant non-uniformities over the surface and significant deviations from ideal blackbody emission), and this, coupled with its not being relevant for any purpose other than the realisation of the candela, meant that relatively few laboratories established an independent scale; even those that did realised the unit on a very infrequent basis and showed rather poor agreement with one another [17]. Secondly, and equally importantly, the relevance of the standard to the types of light source that needed to be calibrated was rapidly eroded. Whereas the colour appearance and luminance of a platinum blackbody is very similar to that of the flame and vacuum filament sources that were used in the first part of the 20th century, these were soon replaced in most lighting applications by more efficient sources, such as gas-filled tungsten filament lamps and fluorescent lamps, which have very different properties. As a result it was necessary to develop methods by which to transfer the unit to these other types of light source, which generally involved visual comparisons and led to rather high measurement uncertainties. The final nail in the coffin for the platinum black-body standard was the development of improved methods for radiometry, paving the way for an alternative definition of the candela that, as well as being easier to implement and more relevant for modern light sources, was also not reliant on setting up a physical artefact but instead could be linked directly to other physical units within the SI. Before examining this new definition, however, it's first necessary to understand the relationships between photometry (the measurement of optical radiation in terms of its visual effect) and radiometry (measurement in terms of optical energy or power).

7.3.1 The relationship between photometry and radiometry

When we refer to 'light', we mean that portion of the electromagnetic spectrum that is directly capable of causing a visual sensation in the human eye, i.e., the narrow band of wavelengths from approximately 360 nm to 830 nm (figure 7.2). It is only to be expected, therefore, that there should be a relatively straightforward relationship between the amount of radiant power reaching the eye and the resulting visual sensation. This relationship can be expressed in the form:

$$X_v = K_m \int_\lambda X_e(\lambda) V(\lambda) d\lambda \tag{7.1}$$

where X_v is the photometric quantity that is to be evaluated (e.g. luminous intensity), $X_e(\lambda)$ is the corresponding radiometric quantity (e.g. spectral radiant intensity) expressed as a function of wavelength, $V(\lambda)$ is the sensitivity of the human eye as a function of wavelength normalised to one at its peak (termed the spectral luminous efficiency function) and K_m is a constant which defines the numerical relationship between photometric units (i.e. the candela in the case of luminous intensity) and the corresponding radiometric units (watt per steradian in the case of radiant intensity).

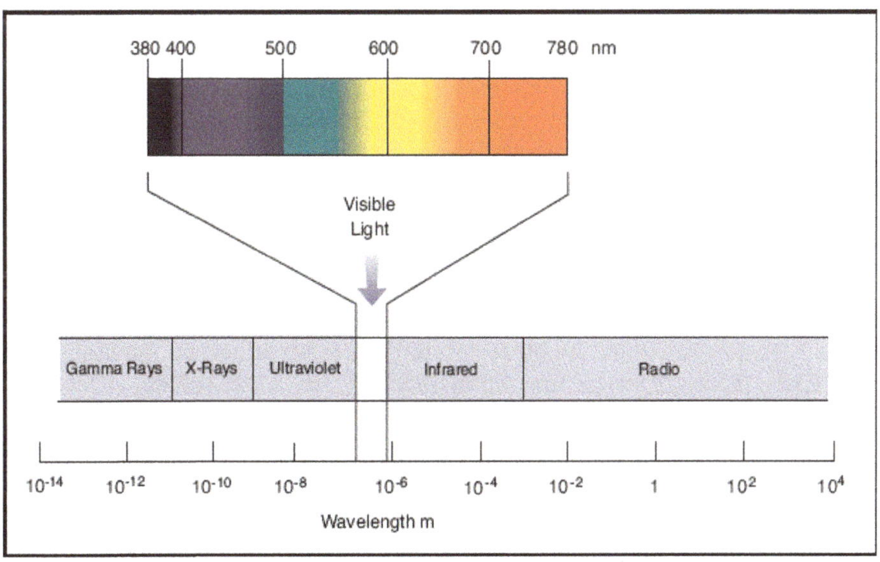

Figure 7.2. 'Light': that portion of the electromagnetic spectrum that is directly capable of causing a visual sensation in the human eye. Courtesy of NPL.

Unfortunately, this relationship is *not* as simple as it first appears. The eye is an extremely complex organ, especially when considered in conjunction with the visual cortex (that part of the brain that processes visual information), meaning that the response of the eye actually varies depending on many factors, such as the lighting level and its spectral (colour) qualities, the position of an observed target in the visual field, and the size of that target. And of course there are further complications arising from the fact that not only does each person see things slightly differently, but for each individual their vision changes with age. As a result, it is not possible to define a single spectral luminous efficiency function that applies for all visual situations; instead all photometry is based on a small number of internationally-agreed spectral luminous efficiency functions that, whilst they do not describe the details of human visual performance, are able to provide a measurement framework for quantifying 'light' in a way that correlates with human vision [18].

The need to define at least one spectral luminous efficiency function arose quite early in the history of photometry. As already described, the definition of the candela (and its predecessors) is based on a specified physical standard and this has a particular distribution of power with wavelength that is defined (albeit indirectly) by the specified operating temperature of the blackbody or the detailed design of the flame standard. If this is used as a reference for calibrating other types of light source, with different spectral characteristics, it is necessary to allow for any differences in the distribution of power with wavelength. Direct visual comparison of light sources of differing colour appearance is difficult and may introduce significant errors, and although these can be minimised by, for example, using coloured filters to adjust both sources to be of similar colour, it soon became preferable to use a photodetector (a device which generates an electrical signal when exposed to light) for such comparisons. However, this approach requires the photodetector to have a response that varies in wavelength in the same way as the human eye, and this in turn requires the response of the eye to be known/defined.

A table of internationally recommended values for the spectral response of the eye was first put forward by the International Commission on Illumination[1] (CIE) in 1924 [19, 20] and adopted without change by the CIPM in 1933. This so-called photopic spectral luminous efficiency function, $V(\lambda)$, characterises the spectral sensitivity of the eye under fully-light adapted (photopic) lighting conditions, for which visual response is governed by the activity of the cones in the retina. The peak of this curve is at 555 nm and the values agreed in 1924 are still used today, essentially unchanged, for the majority of photometric measurements (table 7.1). This has been supplemented by other spectral luminous efficiency functions defined by the CIE [18, 20–23], the first of these being the $V'(\lambda)$ function for scotopic conditions (i.e. conditions under which the eye is fully dark adapted and response is

[1] It should be noted that for more than 75 years the CIPM and the CIE have maintained a special complementary relationship that forms the basis of practical physical photometry. This has been formally recognised in a Memorandum of Understanding and Agreement of Cooperation (signed in April 2007), which states that the CIPM is responsible for the definition of the candela and other photometric units in the SI and the CIE is responsible for the standardisation of the action spectra of the human eye.

dominated by the stimulation of the rods in the retina), which was defined in 1951 [24] and has its peak at 507 nm (see table 7.1).

In the intermediate (mesopic) range between the photopic and scotopic conditions, the rods and cones in the retina are both active, but are stimulated to different degrees depending on the amount of light falling on them, and this leads to a gradual change in spectral sensitivity with changing light level throughout this region. This complex behaviour proved difficult to characterise [26–28], with the result that it was not until 2010 that the CIE published the system for defining the spectral sensitivity functions for use at mesopic levels [29]. This takes the form:

$$M(m)V_{\mathrm{mes};m}(\lambda) = mV(\lambda) + (1 - m)V'(\lambda) \quad \text{for } 0 \leqslant m \leqslant 1 \tag{7.2}$$

where the adaptation coefficient, m, can be determined from the photopic adaptation luminance and spectral characteristics of the visual adaptation field and $M(m)$ is a normalising factor such that $V_{\mathrm{mes};m}(\lambda)$ attains a maximum value of one. If the mesopic luminance of the adaptation field is 5 cd m^{-2} or above, the value of m is one, whereas if the mesopic adaptation luminance is 0.005 cd m^{-2} or below, m is zero. Between these limits the value of m changes gradually, leading to a smooth transition between $V(\lambda)$ and $V'(\lambda)$, as shown in figure 7.3. The spectral characteristics of the adaptation field are expressed in terms of the *S*/*P* ratio, i.e., the ratio of the luminous quantity evaluated according to the CIE scotopic spectral luminous efficiency function to that evaluated using the photopic function; it is the *S*/*P* ratio and the photopic luminance of the adaptation field that determine the value of m. Several other spectral luminous efficiency functions have been defined by the CIE for specific visual conditions, but in practice the vast majority of measurements are made using the photopic, mesopic or scotopic functions, with photopic measurements dominating by far.

7.3.2 Luminous efficacy

The spectral luminous efficiency curves defined by the CIE can be used in equation (7.1) to calculate photometric quantities from the corresponding spectral power measurements on a relative basis, but to obtain absolute values it is also necessary to know the value of the scaling constant, K_m. This scaling constant is termed the maximum spectral luminous efficacy of radiation and is defined for each spectral luminous efficiency function at the peak wavelength of that function; thus there are different maximum spectral luminous efficacies of radiation for photopic vision, scotopic vision, and so on.

7.3.3 The 1979 radiometric definition of the candela

By the 1960s, the difficulties associated with realising the candela using the blackbody definition, coupled with the increasing use of light sources of very different spectral characteristics and the significant improvements that had been made in absolute radiometry, meant that photometric research activity was focused strongly on the feasibility of a radiometric definition of the candela [30, 31]. This generally involved using a spectrally non-selective electrical substitution radiometer (a device

Table 7.1. CIE spectral luminous efficiency functions for photopic vision, $V(\lambda)$, and scotopic vision, $V'(\lambda)$, taken from [25] (quoted at 5 nm intervals and to six decimal places).

Wavelength/nm	$V(\lambda)$	$V'(\lambda)$	Wavelength/nm	$V(\lambda)$	$V'(\lambda)$
360	0.000004	0.000000	600	0.631000	0.033150
365	0.000007	0.000000	605	0.566800	0.023120
370	0.000012	0.000000	610	0.503000	0.015930
375	0.000022	0.000000	615	0.441200	0.010880
380	0.000039	0.000589	620	0.381000	0.007370
385	0.000064	0.001108	625	0.321000	0.004970
390	0.000120	0.002209	630	0.265000	0.003335
395	0.000217	0.004530	635	0.217000	0.002235
400	0.000396	0.009290	640	0.175000	0.001497
405	0.000640	0.018520	645	0.138200	0.001005
410	0.001210	0.034840	650	0.107000	0.000677
415	0.002180	0.060400	655	0.081600	0.000459
420	0.004000	0.096600	660	0.061000	0.000313
425	0.007300	0.143600	665	0.044580	0.000215
430	0.011600	0.199800	670	0.032000	0.000148
435	0.016840	0.262500	675	0.023200	0.000103
440	0.023000	0.328100	680	0.017000	0.000072
445	0.029800	0.393100	685	0.011920	0.000050
450	0.038000	0.455000	690	0.008210	0.000035
455	0.048000	0.513000	695	0.005723	0.000025
460	0.060000	0.567000	700	0.004102	0.000018
465	0.073900	0.620000	705	0.002929	0.000013
470	0.090980	0.676000	710	0.002091	0.000009
475	0.112600	0.734000	715	0.001484	0.000007
480	0.139020	0.793000	720	0.001047	0.000005
485	0.169300	0.851000	725	0.000740	0.000003
490	0.208020	0.904000	730	0.000520	0.000003
495	0.258600	0.949000	735	0.000361	0.000002
500	0.323000	0.982000	740	0.000249	0.000001
505	0.407300	0.998000	745	0.000172	0.000001
510	0.503000	0.997000	750	0.000120	0.000001
515	0.608200	0.975000	755	0.000085	0.000001
520	0.710000	0.935000	760	0.000060	0.000000
525	0.793200	0.880000	765	0.000042	0.000000
530	0.862000	0.811000	770	0.000030	0.000000
535	0.914850	0.733000	775	0.000021	0.000000
540	0.954000	0.650000	780	0.000015	0.000000
545	0.980300	0.564000	785	0.000011	0.000000
550	0.994950	0.481000	790	0.000007	0.000000
555	1.000000	0.402000	795	0.000005	0.000000
560	0.995000	0.328800	800	0.000004	0.000000
565	0.978600	0.263900	805	0.000003	0.000000

570	0.952000	0.207600	810	0.000002	0.000000
575	0.915400	0.160200	815	0.000001	0.000000
580	0.870000	0.121200	820	0.000001	0.000000
585	0.816300	0.089900	825	0.000001	0.000000
590	0.757000	0.065500	830	0.000000	0.000000
595	0.694900	0.046900			

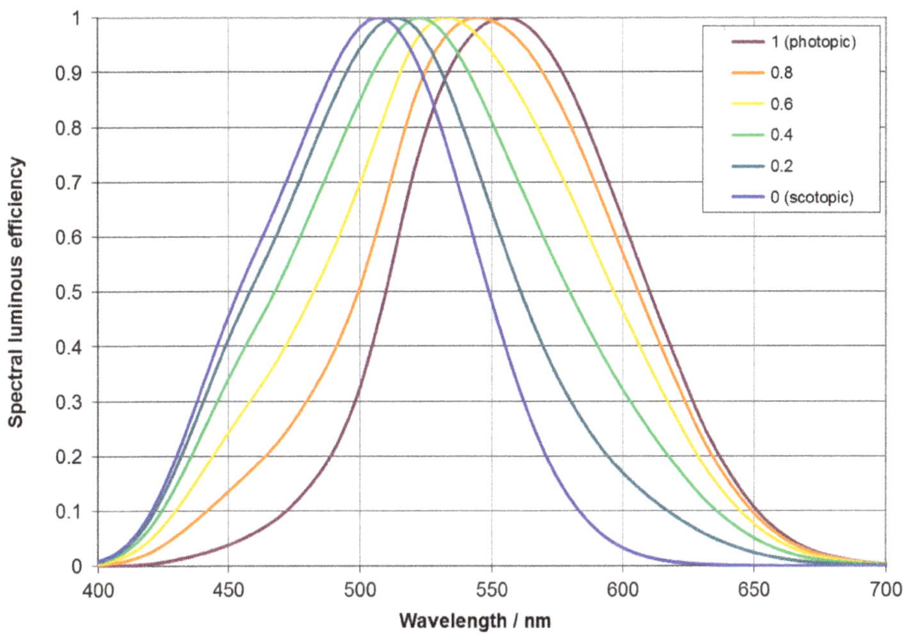

Figure 7.3. Spectral luminous efficiency functions for mesopic vision, for various values of the adaptation coefficient, *m*. Courtesy of NPL.

in which the optical power incident on an absorbing surface is compared with the electrical power required to heat the surface to the same temperature) to measure the radiant flux passing through a precision aperture of known area placed a known distance from a tungsten filament lamp, with a filter placed in front of the radiometer to modify its spectral response to approximate the photopic luminous efficiency function. For the ideal situation, where the response of the filter-radiometer combination exactly matches $V(\lambda)$, the reading of the radiometer, i_v, and the flux through the aperture, $\Phi(\lambda)$, are related as follows:

$$i_v = R_m \int_\lambda \frac{\Phi_e(\lambda)}{d^2} V(\lambda) d\lambda \tag{7.3}$$

where R_m is the absolute responsivity of the filter-radiometer at the wavelength at which $V(\lambda)$ is a maximum and d is the distance between the lamp and the precision aperture. Comparing this with equation (7.1), it is clear that the filter-radiometer signal and the luminous flux, Φ_v (measured in lumen) are also related:

$$i_v = \frac{R_m}{K_m}\frac{\Phi_v}{d^2} \qquad (7.4)$$

Equivalently, since intensity is defined as flux per unit solid angle:

$$i_v = \frac{R_m}{K_m} \cdot \frac{A \cdot I_v}{d^2} \qquad (7.5)$$

where A is the area of the aperture.

Rearranging this gives:

$$I_v = \frac{K_m}{R_m} \cdot i_v \cdot \frac{d^2}{A} \qquad (7.6)$$

Thus if K_m is defined, the filter-radiometer combination gives an alternative method by which to realise the candela; discussions within CCPR during the 1970s were therefore focussed around the best value to choose for this constant [32]. It was agreed that is was highly desirable that the magnitude of the candela should remain unchanged, so several national measurement institutes from around the world carried out experiments to determine the value required to give equivalence with the luminous intensity values assigned to lamps used to maintain their own scale [33]. The mean value obtained from these studies was approximately 683 lm W^{-1} for the maximum luminous efficacy of radiation for photopic vision (as already noted, this is for a wavelength of 555 nm); for scotopic vision a value of $K'_m \sim 1754$ lm W^{-1} was obtained, the peak wavelength being 507 nm in this case.

It was also agreed by the CCPR members that it would be preferable for the value of K_m to be defined at the frequency at which $V(\lambda)$ reaches its peak, rather than the wavelength, to remove the necessity to refer to the refractive index of the medium in which the measurements are being made (although this is usually air, even in this case the refractive index shows changes depending on the atmospheric conditions). Using a refractive index for standard air of 1.000 28, the rounded frequency corresponding to a wavelength of 555 nm is 540×10^{12} Hz (the actual wavelength in standard air for this frequency is 555.016 nm). The final point of debate within CCPR was that although it would be possible to define separate, specific, values for the maximum luminous efficacy of radiation for both of the spectral luminous efficiency functions that had been adopted at that time (i.e. the scotopic function as well as the photopic function), this would be highly undesirable; instead a single value should be used for all situations, allowing a single definition of the candela that holds for all states of visual adaptation [34]. It was quickly realised this could be done by setting the value at the unique wavelength where the numerical relationship between the lumen and the watt is already the same for both scotopic and photopic vision. This wavelength can be found by plotting the spectral luminous efficacy

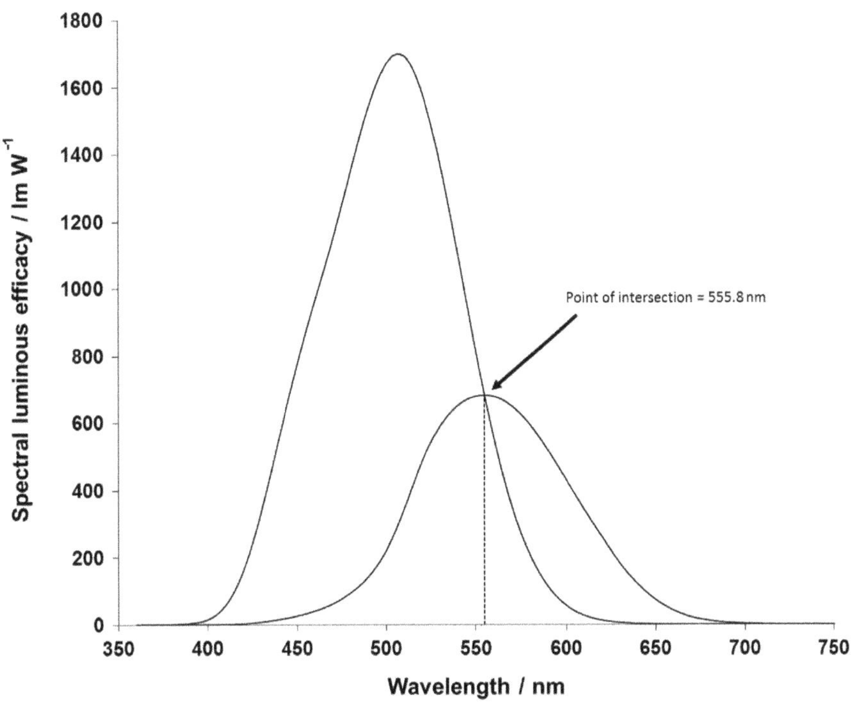

Figure 7.4. Spectral luminous efficacy functions for photopic vision, $K(\lambda)$, and scotopic vision, $K'(\lambda)$ obtained using the 1948 definition of the candela whereby the luminance of a black body (Planckian) radiator at the temperature of solidification of platinum (2042 K according to IPTS-1948) is defined to be 60 cd cm^{-2}. Courtesy of NPL.

functions $K(\lambda) = K_m \; V(\lambda)$ and $K'(\lambda) = K'_m \; V'(\lambda)$, as shown in figure 7.4, and evaluating the wavelength at which they intersect. Using the 1948 definition of the candela, whereby the luminance of a black body (i.e. Planckian) radiator at the temperature of solidification of platinum[2] is defined to be 60 cd cm^{-2}, this point of intersection turns out to be 555.8 nm which, by sheer coincidence, is very close to the wavelength of 555.0 nm at which $V(\lambda)$ reaches a maximum.

As a result of all these debates, the final decision of the CCPR in 1979 was that the value of the luminous efficacy of radiation at a frequency of 540×10^{12} Hz should be defined as 683 lm W^{-1} for all states of visual adaptation. At this frequency the new definition makes $K(\lambda) = K'(\lambda) = 683$ lm W^{-1}, and it follows that, for wavelengths measured in standard air:

$K_m = 683 \text{ lm } W^{-1}/V\,(555.016 \text{ nm}) = 683 \text{ lm } W^{-1}$

and

$K'_m = 683 \text{ lm } W^{-1}/V'\,(555.016 \text{ nm}) = 1700 \text{ lm } W^{-1}$

The new (and still current) definition of the candela agreed by the 16th CGPM of the CIPM in 1979 [35] was:

[2] According to the International Practical Temperature Scale of 1948, IPTS-48, this was 2042 K. Using the current international temperature scale, ITS-90, the temperature is 2045 K.

> The candela is the luminous intensity, in a given direction, of a source that emits monochromatic radiation at a frequency of 540 × 1012 hertz and that has a radiant intensity in that direction of 1/683 watt per steradian.

The consequence of this new definition was that the magnitude of the candela for photopic vision was expected to remain approximately unaltered, whereas for scotopic vision it would change by approximately 3% (this change was considered to be acceptably small, considering the lower accuracy and relative infrequency of scotopic measurements). Subsequent international comparisons largely confirmed this assumption—in the UK, for example, the luminous intensity values assigned to lamps increased by about 0.8% as a consequence of the new definition [36].

7.3.4 Practical photometry using the 1979 radiometric definition

The definition of the candela as given above is expressed in strictly physical terms, independent of any reference artefact, and is stated for only one frequency of electromagnetic radiation[3]. However most light sources emit a radiation over a broad spectrum of frequencies/wavelengths. In practice, therefore, measurements are generally made either by numerical integration of spectral data, using equation (7.1) with K_m set to its defined value of 683 lm W^{-1}, or by comparison with reference sources (usually tungsten filament lamps) which are themselves established according to the definition. Where comparisons are made, the transfer detector used for the comparison is usually a photometer or an illuminance meter (also referred to as a luxmeter), which typically consists of a silicon photodiode coupled with a filter which modifies the spectral response to approximate the $V(\lambda)$ function. An absolutely-calibrated photometer is often also used for realisation of the candela, in a similar manner to the filter-radiometer approach described previously; indeed, this is the method used at the National Physical Laboratory (NPL) to establish the UK's photometric scales [36].

7.3.5 The use of photometric quantities and units

As already discussed, the definition of the candela is independent of the spectral luminous efficiency function that is used to weight radiation at other wavelengths, i.e., it applies for all states of visual adaptation and, more specifically, for all spectral luminous efficiency functions. It is worth noting that it is also implicit in the definition that the spectral luminous efficiency function must obey the law of additivity under the defined conditions for which it applies, and that it must have a non-zero value at a wavelength of 555 nm. These conditions are satisfied for all spectral luminous efficiency functions published by the CIE, which is the only body recognised by CCPR for the standardisation of visual response functions.

Whilst this approach ensures that the definition does not need to be updated each time a new spectral luminous efficiency function is introduced, it also means that the SI units are the same, whichever spectral weighting function is used, which could

[3] See https://www.youtube.com/watch?v=kUwsCCBDAqU

potentially lead to confusion. Luminous intensity, for example, is always measured in candela, and illuminance is always measured in lux, whether measurements are made in the photopic, mesopic or scotopic regime, or under any other defined conditions. Thus in order to avoid any ambiguity, it is essential always to say which spectral luminous efficiency function has been used. This is often done by using descriptor terms such as 'photopic' or 'scotopic', but these must be associated with the quantity name and not, under any circumstances, with the unit; 'scotopic luminous flux' is acceptable, for example, but 'scotopic lumen' is not allowed.

If no spectral luminous efficiency function is specified when expressing a photometric quantity then it is taken, by convention, to be a photopic measurement, i.e., to be evaluated using the $V(\lambda)$ function; for additional clarity, however, the qualifying descriptor 'photopic' may be used. For quantities evaluated using the $V'(\lambda)$ function, the qualifying descriptor 'scotopic' is sufficient, but must always be used. In the case of mesopic quantities the situation is more complicated, since there is no single spectral luminous efficiency function for mesopic vision and hence terms such as mesopic luminance do not uniquely identify the weighting function used. Instead (as described in section 7.3.1) the precise form of the function depends on the visual adaptation conditions and hence these must be described; this is usually done by specifying the value of the adaptation coefficient m.

In summary, photometric values should be expressed in the form of an unambiguous description of the photometric quantity (including identification of the associated spectral luminous efficiency function), a numerical value, and the appropriate photometric unit [37] (see the examples given in table 7.2).

7.3.6 Quantifying other photobiological and photochemical effects

Optical radiation does not only stimulate vision, but is also able to cause changes in a wide variety of living and non-living materials. These changes may occur at the molecular level, whereby one photon interacts with one molecule to change it into new molecular species, or may be thermal in nature, e.g. absorption of infrared radiation in biological tissue can cause damage through tissue coagulation. These interactions between incident optical radiation and the material being irradiated are usually very complicated and are always wavelength dependent. For the purposes of measurement, however, these complexities can be ignored and the effect is characterised simply by means of a weighting function (termed an action spectrum) which describes the ability of monochromatic optical radiation at wavelength λ to produce the given response in the material in question and the associated biological or chemical receptors. Such action spectra are given in relative values, normalised to one at the wavelength at which the efficacy is a maximum; the spectral luminous efficiency functions described previously are specific examples of such action spectra.

For vision, as we have already seen, quantities determined using the relevant action spectra (spectral luminous efficiency functions) are presented using a special SI base unit, the candela. However, this is the only such unit permitted under the SI; in order to avoid a proliferation of units for other photobiological and photochemical effects (or indeed for other biological or chemical effects stimulated by

Table 7.2. Examples of the acceptable use of photometric quantities and units. Note that for those marked *, no qualifying descriptor is used and hence the quantity is assumed to be evaluated for photopic vision i.e. using $V(\lambda)$.

Verbal description	Symbolic description
Photopic luminous intensity of 50.0 cd	$I_v = 50.0$ cd
Luminous flux of 230 lm*	$\Phi_v = 230$ lm
Scotopic luminance of 0.0001 cd m^{-2}	$L\Phi'_v = 0.0001$ cd m^{-2}
Mesopic luminous intensity of 2.3 cd determined using an adaptation coefficient of 0.8	$I_{mes;0.8} = 2.3$ cd m^{-2}

electromagnetic radiation at other wavelengths or by acoustic waves) the SI requires that all other similar effects are quoted in purely physical terms [38].

The procedure for calculating any other photobiological or photochemical effect is very similar to that used for calculating visual effects: the requisite spectral radiant quantity is weighted using the appropriate action spectrum and integrated over the entire spectral range of interest. The difference is that the resultant weighted quantity has the same unit as the spectral quantity and hence when giving a quantitative value, it is essential to specify whether a radiometric or spectrally-weighted quantity is intended, as the unit is the same. As an example, an assessment of the photobiological safety of a light source requires determination of the retinal blue light hazard [39, 40] (the potential for a photochemical-induced retinal injury resulting from exposure at wavelengths primarily between 400 nm and 500 nm), which in turn requires the blue light hazard weighted radiance, L_B, to be calculated from the measured spectral radiance distribution $L(\lambda)$. This is done by weighting the spectral radiance of the source at wavelength λ by the blue light hazard function $B(\lambda)$ and summing over all wavelengths present in the source spectrum over the full wavelength range of this action spectrum (300 nm to 700 nm):

$$L_B = \int_{300}^{700} L(\lambda)B(\lambda)\mathrm{d}\lambda \tag{7.7}$$

Since $L(\lambda)$ is expressed in W m^{-2} sr^{-1}, so too is L_B. Thus when quoting the result, the weighting function used (the blue light hazard function in this case) must also be explicitly stated.

7.4 A look to the future

The current definition of the candela has stood the test of time well. Although improvements have been made to the techniques used to realise the candela, particularly in terms of the performance of the radiometers on which such realisations are typically based, no change to the definition itself has been necessary. It has met the objective of allowing the introduction of new spectral luminous efficiency functions, such as those defined for mesopic conditions, without any change being required to the unit or its definition, and continues to provide a consistent and reliable basis for the evaluation of optical radiation in a way that correlates well with its visual effect for all states of visual adaptation. The uncertainties that can be achieved for the realisation of the candela using the present definition are around 0.2% (at the 95% confidence level) and although this appears high when compared with most of the other SI base units, it is generally both adequate and appropriate for the purposes of photometry: in practice final photometric measurement uncertainties for the purposes of product specification, quality control, design of lighting installations etc., are only required at the few percent level. It therefore appears unlikely that any major change to the definition will be made in the short term, although it is expected that it will be reformulated in 2018 so it can be expressed in terms of a defined constant, so as to be consistent with revised definitions for the other base units that are being introduced at the same time. The reformulated definition is expected to be:

> The candela, symbol cd, is the SI unit of luminous intensity in a given direction. It is implicitly defined by taking the fixed numerical value of the luminous efficacy of monochromatic radiation of frequency 540×10^{12} Hz, K_{cd}, to be 683 when expressed in the unit lm W^{-1}, which is equal to cd sr W^{-1}, or cd sr kg^{-1} m^{-2} s^{3}, where the kilogram, metre and second are defined in terms of h, c and $\Delta\nu_{Cs}$.
>
> The effect of this definition is that one candela is the luminous intensity, in a given direction, of a source that emits monochromatic radiation of frequency 540×10^{12} Hz and has a radiant intensity in that direction of (1/683) W sr^{-1}.

The success of the present definition does not mean, however, that research into potential future improvements and developments has ceased. On the contrary, National Measurement Institutes (NMIs) have continued to investigate and exploit advances in detector and source technology in order to reduce calibration uncertainties and improve the ease with which photometric (and radiometric) scales can be realised and disseminated [41]. Most NMIs now establish the candela by radiometric methods, using one of two approaches:

- A reference photometer (a filter-detector combination with a spectral responsivity which provides a reasonably close approximation to the $V(\lambda)$ function) is calibrated in terms of its relative spectral responsivity over its full wavelength range, with its absolute irradiance responsivity determined at one (or more) wavelengths against a high accuracy radiometer. The absolutely-calibrated photometer can then be used as a reference artefact for direct calibration of the illuminance (or the luminous intensity if the source-photometer distance is known) of other sources, provided that appropriate corrections are applied to allow for any mismatch between its actual spectral responsivity and the $V(\lambda)$ function. Alternatively, the photometer can be used at a known distance from a transfer standard lamp (usually a tungsten filament lamp operating at a correlated colour temperature of around 2856 K and set up in a specified geometric configuration), allowing the luminous intensity of this lamp to be calibrated and used as a reference for calibration of other sources. Using this ‘reference photometer’ approach, tungsten filament standard lamps can be calibrated with expanded uncertainties for luminous intensity measurements as low as 0.2% [36, 42–44].

- A reference polychromatic source (commonly a tungsten filament lamp) is measured in terms of its spectral radiant intensity at a few discrete wavelengths in the visible wavelength range using either a series of narrow-band, absolutely calibrated, filter-radiometers or an absolutely calibrated spectroradiometer. The measured radiant intensity of the polychromatic source is then interpolated and extrapolated to other wavelengths as required, using a model of the typical spectral behaviour of that source, to give values over the entire visible wavelength range; these values can then be weighted by the $V(\lambda)$ function and spectrally integrated to give the corresponding luminous intensity [45]. Uncertainties using this approach are higher than can be achieved using the reference photometer method, due to the additional uncertainties associated with the interpolation of the spectral radiant intensity values, but are still acceptably low for most applications.

Both of these approaches require absolute calibration of a detector at some point in the realisation. Often this is a cryogenic cavity radiometer [46, 47], a form of electrical substitution radiometer in which the measurement uncertainties are minimised through the use of a highly absorbing cavity coupled with operation at liquid helium temperatures. Using this device, radiant power measurements can be made with an expanded uncertainty of just a few parts in 10^5. However cryogenic radiometers are not only expensive, but can be difficult to use, and much recent research activity in the area of radiometry has therefore focussed on the development of potential alternatives. So-called predictable quantum efficient photodiodes (PQEDs) [48, 49] have proved particularly successful: these are based on a low loss semiconductor, generally silicon, together with an accurate model of the photon-to-electron conversion and detection within the device, which allows the incident optical radiation to be determined from measurement of the photocurrent generated. This method was initially applied to single photodiodes [50] but more recently it has been used for a number of photodiodes arranged in a ‘trap’ configuration, in which the multiple reflections within the trap not only increase the overall detection efficiency, but also improve the reliability of the model. Such PQED devices now allow standard uncertainties as low as 0.01% to be achieved in absolute spectral radiant power or irradiance measurements, at a fraction of the cost of a cryogenic radiometer. The use of PQEDs for direct determination of the luminous intensity or illuminance of relatively narrow-band sources, such as coloured LEDs, has also been demonstrated [51]; this approach has the advantage that it does not require an intermediate tungsten filament lamp to hold the luminous intensity scale, but the disadvantage that the relative spectral irradiance as a function of wavelength must be measured in a separate experiment.

An alternative method by which to realise radiometric and photometric scales is to use a primary standard source and this, too, continues to be an area of active research. The most commonly used primary sources for radiometric purposes are blackbody radiators and electron storage rings [52, 53], both of which produce radiant power over a wide spectral range with an output as a function of wavelength that can be calculated from fundamental principles from knowledge of their key physical parameters. Of course the 1948 definition of the candela was based on the use of blackbody, but the major advance that has been made during recent years is the development and application of ultrahigh temperature blackbody sources, able to operate at temperatures in excess of 3000 K [54]. If the temperature of such a blackbody is known (usually determined by measuring its absolute radiance over a narrow band of wavelengths), then its spectral radiance as a function of wavelength is also known via Planck's radiation law:

$$L(\lambda, T) = \frac{2hc^2}{\lambda^5}\left(\frac{1}{\mathrm{e}^{hc/\lambda \mathrm{k}T} - 1}\right) \tag{7.8}$$

where k is Boltzmann's constant (1.3807×10^{-23} J K^{-1}) and T is the thermodynamic temperature of the blackbody in kelvin. These high temperature blackbodies are most typically used in the establishment of scales of spectral radiance and irradiance [55, 56] but luminous intensity can also be obtained by weighting the spectral radiant intensity with the spectral luminous efficiency function. Since the best uncertainties for spectral radiance and irradiance using a high temperature blackbody are typically about 0.2% to 0.3% (at the 95% confidence level) in the visible spectral region, the uncertainty for luminous intensity realised using this approach is also around this level; uncertainties using a synchrotron source in the visible region are considerably higher and these are therefore not used for realisation of photometric scales.

The final major area of research that may offer a new approach for radiometry and photometry is 'quantum' or photon counting techniques. These are being developed primarily for quantum optics, quantum computing, communication, security and similar applications, and the uncertainties that can currently be achieved are considerably higher than with classical radiometric methods. However, rapid technological developments are underway and it appears feasible that it will be possible within a few years to produce a radiant flux at photon counting levels with an uncertainty approaching, or potentially even better than, that offered by more traditional methods (see [41] and the numerous references therein). It is interesting to note that the recent *mise en pratique* for the candela [57] already includes information relating to the use of photon counting techniques for the establishment of radiometric and photometric scales, and these methods can be used without any need for a change to the definition of the candela.

References

[1] 1877 *J. Gas Lighting* **30** 337

[2] 1897 *J. Gas Lighting* **70** 1050

[3] 1884 *Electrician* 12 511

[4] Draper J W 1880 *Philos. Mag.* **9** 76

[5] Violle J 1884 *Philos. Mag.* **17** 563

[6] Petavel J E 1899 *Proc. R. Soc.* **65** 469
[7] 1909 *Illumin. Eng.* **2** 393
[8] 1921 Recueil des travaux et compte rendu des séances CIE 5th session, held in Paris
[9] Wensel H T, Roeser W F, Barbrow L E and Caldwell F R 1931 *J. Res. Bur. Stand.* **6** 1103
[10] Buckley H and Barnett W 1937 Procès-verbaux des séances *CIPM* **18** 247
[11] Ribaud G 1933 *Rev. d'Optique* **12** 289
[12] 1937 Procès-verbaux des séances *CIPM* **18** 236
[13] 1940 National Physical Laboratory Report for the year 1939 (London: HMSO) 37
[14] 1946 Procès-verbaux des séances *CIPM* **20** 119
[15] 1948 Recueil des travaux et compte rendu des séances CIE 11th session, held in Paris
[16] 1968 *Metrologia* **4** 43–4
[17] Bonhoure J 1971 *BIPM CCPR 7 Annexe* **P12**
[18] 1978 CIE Publication **41** 1978
[19] 1924 Recueil des travaux et compte rendu des séances CIE 6th session held in Geneva
[20] Preston J S 1940 *Trans. Illum. Eng. Soc* **5** 109
[21] 1978 CIE 41:1978 2006 ISO 11664-1:2007(E)/ CIE S 014-1/E:2006
[22] 1988 CIE Publication **75** 1988
[23] 1990 CIE Publication **86** 1990
[24] 1951 CIE Compte Rendu **3** Table II pp 37–9
[25] 2004 ISO 23539:2005(E)/CIE S 010/E:2004
[26] 2001 CIE Publication **141** 2001
[27] Rea M S *et al* 2004 *Lighting Res. Technol.* **36** 85
[28] Goodman T M *et al* 2007 *Lighting Res. Technol.* **39** 365
[29] 2010 CIE Publication **191** 2010
[30] Preston J S 1963 *Proc. R. Soc. London Ser.* A **272** 133–45
[31] Gillham E J 1963 *Proc. R. Soc. London Ser.* A **278** 137–45
[32] Blevin W R and Steiner B 1975 *Metrologia* **11** 97
[33] 1977 BIPM CCPR 9 P6-P8
[34] Blevin W R 1979 CIE Proc. P-79-02
[35] 1980 *Metrologia* **16** 56
[36] Goodman T M and Key P J 1988 *Metrologia* **25** 29–40
[37] 2016 CIE TN 004:2016
[38] 2016 BIPM The International system of units (SI) 8th Edition Appendix 3
[39] 2002 IEC 62471:2006/CIE S 009:2002
[40] 2014 CIE TN 002:2014
[41] Zwinkels J C *et al* 2010 *Metrologia* **47** R15
[42] Cromer C L *et al* 1996 *Res. Natl Inst. Stand. Technol* **101** 109
[43] Erb W and Sauter G 1997 *Metrologia* **34** 115
[44] Toivanen P *et al* 2000 *Metrologia* **37** 131
[45] Kärhä P *et al* 1997 *Appl. Opt.* **36** 8909
[46] Martin J E, Fox N P and Key P J 1985 *Metrologia* **21** 147–55
[47] Hoyt C C and Foukal P V 1991 *Metrologia* **28** 163–7
[48] Sildoja M *et al* 2013 *Metrologia* **50** 385–94
[49] Müller I *et al* 2013 *Metrologia* **50** 395–401
[50] Geist J 1979 *Appl. Opt.* **18** 760
[51] Dönsberg T *et al* 2014 *Metrologia* 51 S276

[52] Klein R *et al* 2009 *Metrologia* 46 S266
[53] Arp U *et al* 2000 *Metrologia* **37** 357
[54] Sapritsky V I *et al* 1997 *Appl. Opt.* **36** 5403
[55] Metzdorf J 1993 *Metrologia* **30** 403–8
[56] Yoon H W, Gibson C E and Barnes P Y 2002 *Appl. Opt.* **41** 5879
[57] Zwinkels J C *et al* 2016 *Metrologia* **53** G1

IOP Publishing

Precise Dimensions
A history of units from 1791-2018
Malcolm Cooper and Jim Grozier

Chapter 8

The story of mass standards 1791–2018[1]

Jim Grozier, Sally Riordan and Richard Davis

8.1 Introduction

This story of mass standards begins with an attempt to abolish primary mass standards in the late 18th century and ends with another attempt to abolish primary mass standards in the early 21st. A *standard*, in this context, is a physical artefact which defines a unit—it is the physical embodiment of that unit. A *primary* standard is one that defines the unit without reference to anything else.

We have not attempted a global survey and instead have concentrated mainly on French and British standards. We cover *international* standards only in the sense that, following the Metre Convention and the creation of the BIPM (*Bureau International des Poids et Mesures*), the standards maintained by that body are held to be international.

We first describe the genesis of the Kilogram of the Archives (KA) the first standard kilogram (section 8.2) and then cross the channel to review early English weight standards (section 8.3). Section 8.4 deals with the International Prototype Kilogram (IPK), the second standard, which replaced the KA in 1889. We investigate the stability of physical standards in section 8.5, leading to the decision to redefine the kilogram (section 8.6). Finally (section 8.7) we take a look at one of the two proposed methods for realising the new kilogram—the silicon x-ray crystal density method.

8.2 Construction of the kilogram of the archives

Antoine-Laurent Lavoisier (1743–1794) was a chemist, best known today for his rôle in what has become known as the Chemical Revolution, in which a new model of

[1] This chapter is based on a talk given by Richard Davis at the meeting *A History of Units from 1791 to 2018*, held at NPL on 17 March 2016. For additional background material on section 8.2, see Riordan (2013), Riordan (2015) and Davis *et al* (2016); for additional background material on sections 4–7, see Davis *et al* (2016).

doi:10.1088/978-0-7503-1487-9ch8 © IOP Publishing Ltd 2017

combustion was adopted, which saw it as combination with oxygen rather than the release of phlogiston. As the astronomers Méchain and Delambre set about measuring the meridian from Dunquerque to Barcelona, Lavoisier and the mineralogist Abbé René-Just Haüy (1743–1822) were given the task of creating the new mass unit and calibrating it against the old standard, the Pile de Charlemagne. Both were members of the Commission des Poids et Mesures, a subgroup of the Académie des Sciences.

The new unit, the *grave* (later renamed the kilogram), as adopted by the French Assembly in 1791, was defined to be the mass of one cubic decimetre of distilled water at the melting point of ice. Thus, like the new unit of length, the mass unit was to be based on a quantity 'taken from nature'—the density of water, which, together with the metre, yields a mass.

The definition of the *grave* lent itself to a particularly simple realisation procedure. A solid body with the same density as distilled water at the melting point of ice would float in it. If such a body were constructed so as to have a thin stalk protruding from its upper face, its overall density being slightly *less* than that of freezing distilled water, the density of the body only (without the stalk) being slightly *greater*, it would float in such a way that the body itself was completely submerged but the stalk protruded from the surface of the water. If a mark were made on the stalk at a point below which the volume of the body was exactly one cubic decimetre, the mark would be at surface level if the total mass of the body were one *grave* (and proportionately if the volume were not one cubic decimetre). If it were above the surface, small weights could be added to lower it into position.

Lavoisier and Haüy commissioned such a body, in the form of a hollow copper cylinder with a thin stalk attached; it was constructed by the instrument-maker Nicolas Fortin. They made a mark corresponding to a known volume and floated it in ice-cold water. The mass of the body was then determined in air against a set of provisional mass standards.

A number of corrections were applied, including a buoyancy correction for the weighing in air and an allowance for the fact that they did not have time to distil enough water and instead used filtered water from the Seine, which was slightly more dense and was not at freezing point. The reason for their haste was because the French mint needed to manufacture the new currency, the *franc*, which had been defined as one *centigrave* of silver. This required a calibration between the *grave* and the *livre* and *grain*, the old standards on which the Pile de Charlemagne was based, where one *livre* equalled 9216 *grains*[2].

No details of the sizes of the individual corrections have survived, but it is possible to estimate them from the data available at the time. One of us (SR) has investigated this process in some detail[3]. It is worth summarising it here, especially when one

[2] The grain derives from the use of seeds as units of weight; Connor tells us that these were the earliest units. However, different countries used different grains: wheat grains in medieval France, barleycorns in England. The carob seed was the basis of the *carat*, a weight still in use for precious metals. Connor (1987) p 2.

[3] See Riordan (2013) pp 32–68.

Figure 8.1. The Pile de Charlemagne. © Musée des Arts et Métiers, CNAM, Paris.

considers that an eminent historian of science, C Gillispie, has claimed that Lavoisier and Haüy 'had taken no account of physical variables of temperature and atmospheric pressure'[4]. In Box 1 we summarise what was probably done, albeit in a modern algebraic format that would not have been used in the 1790s.

Box 1: Lavoisier and Haüy's Determination of the Grave

Figure 8.2 shows the cylinder, mass m_C and volume V_C, floating in water (density ρ_H):

$$m_C g = V_C \rho_H g \;\Rightarrow\; m_C = V_C \rho_H \tag{8.1}$$

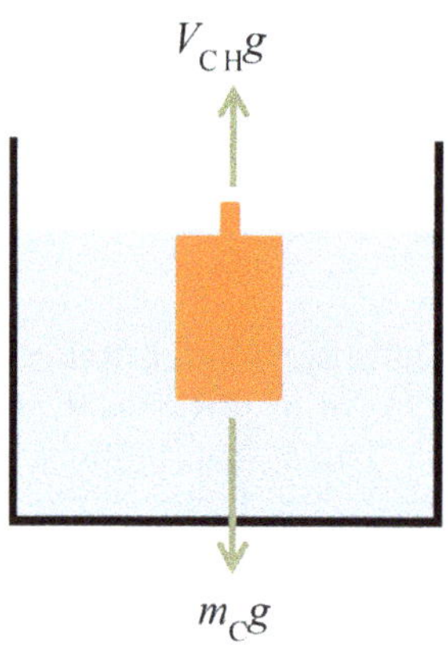

Figure 8.2. The cylinder in water.

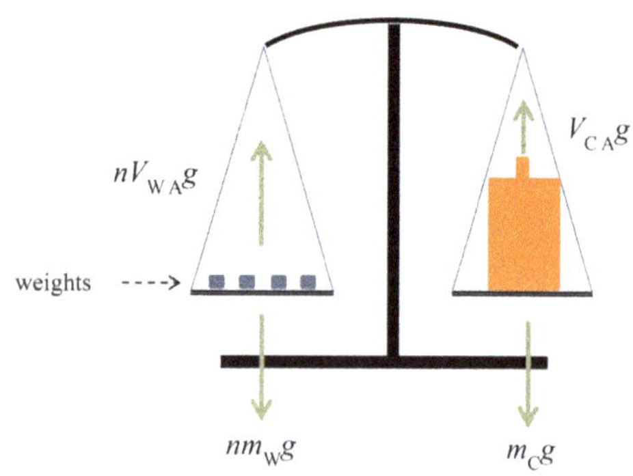

Figure 8.3. The cylinder on the balance.

[4] Gillispie (2004) p 472.

Figure 8.3 shows the cylinder balanced by n identical weights, each of mass m_W and volume V_W and acted on by two buoyancy forces: [5]

$$nm_W g - nV_W \rho_A g = m_C g - V_C \rho_A g \Rightarrow nm_W - nV_W \rho_A = m_C - V_C \rho_A \quad (8.2)$$

where ρ_A= density of air. Substituting (1) into (2):

$$\begin{aligned} nm_W - nV_W \rho_A &= V_C \rho_H - V_C \rho_A \\ \therefore \quad \frac{nm_W - nV_W \rho_A + V_C \rho_A}{V_C} &= \rho_H \end{aligned}$$

Factors a,b,c,d were applied to correct for:

a: river water being denser than distilled water; b: the water not being at freezing point; c: the copper rulers used to measure the cylinder against the Fathom of Peru not being at the latter's defining temperature; d: the cylinder being at a slightly higher temperature than when it was measured. Hence:

$$\begin{aligned} \frac{nm_W - nV_W \rho_A + V_C \rho_A}{cdV_C} &= ab\,\rho_H \\ \therefore \quad m_W &= \frac{abcd\rho_H V_C + nV_W \rho_A - V_C \rho_A}{n} \end{aligned} \quad (8.3)$$

Here, ρ_H is, by definition, one *grave* per cubic decimetre. The weights can then be calibrated against the old mass scale to give the conversion factor between *grave* and *grain*.

[5] It is not totally clear whether the cylinder was open to the air, or closed. If the former, there would be a much smaller buoyancy force on the balance, but there would be an extra term representing the weight of contained air when the cylinder was in the water. Whichever is the case, the corrections are numerically identical.

V_C was calculated from careful measurements of the cylinder; V_W was also presumably known, although the buoyancy force on the weights would have been much smaller than that on the cylinder and may have been ignored. ρ_H was, by definition, 1 *grave* per cubic decimetre. From a knowledge, or at least an estimate, of a,b,c,d and ρ_A, therefore, m_W could be found in *graves* and a further weighing of the weights against the Pile de Charlemagne would hence deliver the required conversion factor between the *grave* and the *grain*[6]. Additionally, it was necessary to convert the measurements of the cylinder into metres using a provisional conversion factor, since the metre had not yet been realised in terms of its definition.

The result of this preliminary weighing was that one *grave* was equivalent to 18 841 *grains*. This calibration was accepted by the French assembly on 1 August 1793. By April 1795, the *grave* had been renamed the *kilogram*.

[6] In theory, the temperature correction should also be applied to the buoyancy term, but this would be a second order correction.

Lavoisier and Haüy had plans for the completion of their task, including a more accurate comparison of the density of their filtered water with that of distilled water when they had a sufficient amount and also building an oxygen blast furnace to refine platinum. But the work was never completed; in 1794 the Revolution caught up with them—Lavoisier was guillotined because of his rôle as a tax collector and Haüy—who was considered a 'recalcitrant priest'—was only saved from the guillotine by friends. A new team took over the task of refining the *grave*, consisting of Louis Lefèvre-Gineau (a former student of Lavoisier) assisted by Giovanni Fabbroni and once again by Fortin, who constructed a new hollow cylinder, slightly more dense than water[7]. This was weighed in air and then in water—this time at its *maximum density* (at about 4 °C) rather than at the temperature of melting ice. H W Chisholm notes that 'it was not thought requisite that the cylinder should be of the specified volume of a cubic decimetre, but only of the most convenient size for arriving at the desired result by computation'[8]. The cylinder was actually just over 11 cubic decimetres; scaling up by this factor meant that the fractional uncertainty in the weight would reduce correspondingly, facilitating more precise measurements. In fact, Lavoisier and Haüy's cylinder had been of a similar size[9].

Having made precise measurements of the volume of the cylinder, they were able to derive a value for the upthrust, or buoyant force, on the cylinder when immersed, which, by the definition of the kilogram, was simply the same number of kilograms as the volume in cubic decimetres. The results of the weighings then enabled them to compare this with the weight of Lavoisier and Haüy's provisional kilogram, by means of a set of copper weights, newly constructed for the experiment, but based on their predecessors' preliminary value of 18 841 *grains* to the *grave*[10]. They found that this provisional kilogram was slightly too heavy and revised the conversion factor between old and new measures to 18 827.15 *grains* per kilogram[11]. As Trallès recounted, the apparent mass of the hollow cylinder in water was in need of many corrections. He explained these very clearly; they are similar to the corrections needed today for the determination of volume by hydrostatic weighing.

The team fashioned a new standard, made of platinum sponge, using platinum refined by Marc-Étienne Janety, a goldsmith by trade. This became known as the *Kilogram of the Archives* (KA). It was declared to be the 'definitive standard of mass' in a French law passed on 10 December 1799. Whereas Fortin, who was a gifted

[7] There is some confusion regarding the material, not only of Lefèvre–Gineau's cylinder but also of Lavoisier's and of the Pile de Charlemagne, with some scholars referring to *cuivre* (copper) and some to *laiton* or *cuivre jaune* (brass). Trallès and Delambre both describe Lefèvre–Gineau's cylinder as composed of *laiton*; however, *cuivre jaune* was the common word for brass until the 1950s and in musical ensembles, 'les cuivres' denotes the *brass* section (Suzanne Hinton, personal communication to JG, 12.4.17). See also Riordan (2013) p 158 for a discussion of this question.

[8] Chisholm (1873) p 388. See also Riordan (2013) p 55.

[9] Riordan (2015) p 39.

[10] Riordan (2013) p 162.

[11] Connor (1987) p 353.

Figure 8.4. The International Prototype Kilogram. © BIPM, Sèvres.

instrument maker, is referred to as an *artiste* in the official reports, Janety was a tradesman whose contribution is not mentioned.

In 1804, Napoleon became Emperor. He was under pressure to restore the old measures and in a decree of 1812, old unit names and non-decimal subdivisions were allowed, whilst the standards themselves remained unaltered. Napoleon died in 1821, but the original metric system was not restored until a law passed in 1837, which McGreevy says was the 'real beginning of the full metric system in France'[12].

8.3 William Hallowes Miller and the New Imperial Standard Pound

After the Revolution, the British Government 'wished to keep anything and everything French at a distance lest the rebellious contagion spread'[13]. Nevertheless, the simplicity of the metric system did attract admirers in Britain. Part of the attraction was the way the metric units were linked together; as Connor

[12] McGreevy (1995) pp 149–50.
[13] Connor (1987) p 279.

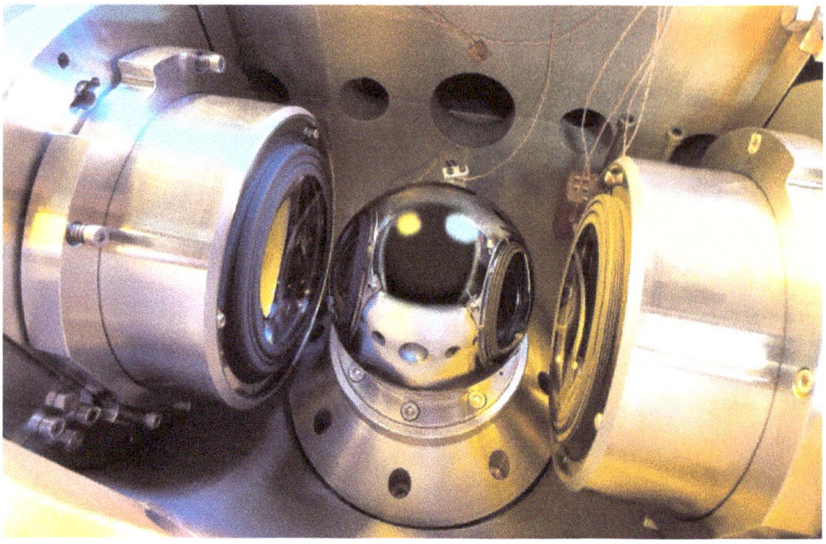

Figure 8.5. Silicon sphere in a sphere interferometer. Photo courtesy of Physikalisch-technische Bundesanstalt (PTB), Braunschweig, Germany.

puts it, 'the metre defined the kilogram which in turn defined the litre'[14]. There was also considerable support for decimal currency, again following the French lead. The possibility of decimalising the British coinage was raised in the House of Commons in 1816 and again in 1847, 1853 and 1855 and various commissions established. The only concrete outcome of this, however, was the issue of a coin worth 1/10 of a pound sterling, called the *florin*[15].

A Commons debate on weights and measures in 1816 fared slightly better; it resulted in a comparison being made by Captain Henry Kater, on behalf of the Royal Society, between the metre and the Imperial standard yard. Subsequently, a Royal Commission was set up 'to consider how far it might be practicable or advisable to establish a more uniform system of weights and measures'. Up to that point, English weights had been anything but uniform, due partly to the coexistence at various times of two or more different versions of the pound.

William Hallowes Miller's paper *On the Construction of the New Imperial Standard Pound and its Copies of Platinum; and on the Comparison of the Imperial Standard Pound with the Kilogramme des Archives* traces the history of English weight standards from before the Norman Conquest. Miller states that 'the earliest legal standard of English weight, of which any very authentic account is preserved, is the weight called the pound of the Tower of London'. This weight is also referred to

[14] Connor (1987) is not entirely consistent on this point, however. He tells us (p 279) that the litre was originally identical to a cubic decimetre—so, not dependent on the kilogram—until its redefinition in 1901 following the discovery that the volume of a kilogram (defined in terms of the IPK) of water at 4 °C was not exactly one cubic decimetre but in fact 1.000 028 dm^3 (pp 354–5). This redefinition was abrogated in 1964. He gets the sequence right on the previous page, however.
[15] Connor (1987) pp 281–2.

as the *tower pound.* This standard was, however, superseded by the slightly heavier Troy Pound; a statute made in the year 1498 confirmed the use of the troy pound as a legal standard and a later document dated 1527 decreed that 'It is determyned by the King's highness and his said councelle, that the foresaid pounde Towre shall be no more used and occupied, but al maner of golde and sylver shall be wayed by the pounde Troye ...'[16].

Somewhat heavier than the troy pound is the *avoirdupois* pound, which Miller traces as far back as the time of Edward III, although he adds that 'it is not known when the avoirdupois weight was first introduced'. McGreevy points out that this word, which he suggests derives from the Latin *averium ponderis*, referred originally to a class of goods rather than a weight, the weight itself being the mercantile pound and the goods for which it was used including 'wax, sugar, pepper, almonds' and some herbs; 'but the most important ... was wool'[17].

In 1758 a Committee of the House of Commons 'recommended that the troy pound should be made the unit or standard by which the avoirdupois and other weights should be regulated'. The Committee's report refers to three troy pounds which 'were adjusted with great care'; one of these was 'placed in the custody of the Clerk of the House of Commons' while the others are thought to have been given to a Mr Harris, the Assay-Master and a Mr Freeman, 'weight-maker to the Mint, the Exchequer and the Bank of England'[18]. The resulting bill did not complete its passage through Parliament, due to the death of King George II.

The Royal Commission's recommendations became law in an Act of Parliament in 1824. These included the observation that 'the avoirdupois pound, which has long been in general use, though not established by any act of the Legislature, is so nearly 7000 troy grains, that they recommend that 7000 such troy grains be declared to constitute a pound avoirdupois'. The Commission further determined, from weighings done by Sir George Shuckburgh in 1797 and linear measurements by Captain Kater in 1821, 'the weight of a cubic inch of distilled water, weighed in air by brass weights, at the temperature of 62 °F, the barometer being at 30 inches, to be equal to 252.458 grains, of which the Imperial standard troy pound contains 5760'. The standard troy pound which had been kept by the Clerk of the House of Commons since 1758 'shall continue to be the original unit or only standard of weight from which all other weights shall be derived'; it was to be known as 'The Imperial Standard Troy Pound'[19]. The calibration of the troy pound in terms of the density of water at a given temperature and pressure allowed the 1824 Act to also incorporate a clause for the reconstitution of the standard if it should be lost or destroyed, by reference to the weight of a cubic inch of water.

Miller reports that, when the Houses of Parliament burned down in 1834, 'all the standards of measure and weight were either totally destroyed, or injured to such an

[16] Miller (1856) pp 753–4.

[17] McGreevy (1995) p 67. He quotes R E Zupko as listing 27 variations for the spelling of 'the phrase currently expressed by the single word "avoirdupois"'.

[18] Miller (1856) pp 757, 793.

[19] Miller (1856) pp 757, 793.

extent as to render them quite useless as standards'. This 'all' presumably refers to all those standards that were actually kept there, but does *not* include other standards, including the two remaining troy pounds that had been referred to in the 1758 report; these are thought to have been among the secondary standards that Miller later compared[20].

In the wake of the fire, a Commission was set up to decide on how the standards could be replaced. It reported in 1841 that it had decided against adopting the method laid down in the 1824 Act for reconstituting the standard pound using the density of water as standard, because 'the determination of the weight of a cubic inch of water is yet doubtful'; on the other hand, 'several metallic weights exist, which were most accurately compared with the former standard pound'. It is not clear from Miller's account whether these included the two standards that had been 'adjusted with great care' in 1758. Furthermore, the Commission was confident that standards could always be more securely restored by using 'material copies which have been carefully compared with them' than 'by reference to experiments referring to natural constants'[21].

The 1841 report also observed that 'the avoirdupois pound, being universally used through the kingdom, while the troy pound is wholly unknown to the great mass of the population, be adopted as the standard of weight'[22]. Use of the troy pound was henceforth to be restricted to 'gold, silver and precious stones' and it has continued in this rôle to the present day. In 1844, in pursuance of this new standard, five 'platinum weights, a little in excess of 7000 grains' were made by a Mr Barrow[23].

In 1843, Miller himself was appointed a member of the Committee which arose out of the 1841 report and was personally charged with the construction of the new standards of weight. First of all, he gathered together a number of secondary standard troy pounds and compared their weights. He then refers to the work of a Professor Moll, who in 1831 had compared standard troy pounds with a standard kilogram that had been presented to the British Government by the Committee of Weights and Measures in Paris, together with several other standard kilograms that he had acquired. Moll's assessment of this exercise was that 'the result has left me in an entire darkness as to the real value of the kilogramme'[24].

Miller travelled to Paris in 1844 in order to compare a standard pound with the Kilogram of the Archives. He reported that 'by some most unaccountable oversight (the Kilogram of the Archives) had never been weighed in water previous to its final adjustment. Afterwards, on account of its legal importance, it was considered hazardous to immerse it in water' because of the possible presence on the surface of arsenic which might dissolve in the water 'and thus produce a very sensible alteration of weight'[25]. Arsenic had been used in the refining process by Janety when he

[20] Miller (1856) pp 759, 793.
[21] Miller (1856) pp 759, 793.
[22] Miller (1856) p 761.
[23] Miller (1856) p 806.
[24] Miller (1856) p 874.
[25] Miller (1856) p 875.

constructed the standard. In the mid-18th century the facilities for melting platinum by itself did not exist; instead, Heinrich Scheffer of the Swedish mint, who is credited with the first serious investigations of the element, found that it melted 'in the twinkling of an eye' if mixed with a little arsenic[26].

Weighing the standard in both water and air was done to determine the volume of the standard, where the masses and volumes of the balance weights must be known as well as the density of water and air. The volume so measured was needed to correct for the effect of air buoyancy when the standard was subsequently used in air to calibrate secondary standards of known volume. Miller noted that the balance he used for comparing standards had a facility for doing this: the base 'has an opening immediately under the right-hand pan' and 'a corresponding opening in the table on which the balance stands. The vessel of water is placed under the table ...'[27].

Because of the problems associated with immersing it in water, he instead used a *stereometer* (a gas pycnometer, where air is the gas) to compare the volume of the Kilogram of the Archives with that of a brass cylinder of similar size, which could afterwards be weighed in water[28]. He found that the density of the standard was 20 549 kg m^{-3}; since the density of pure platinum was known to be about 21 460 kg m^{-3}, the defect of about 4% was assumed to be due to either residual arsenic or holes in the platinum.

Miller's value for the pound in terms of the KA was that 1 lb equalled 453.592 652 5 grams. Several decades later, in 1883, the pound and kilogram were again compared, the result being 453.592 428 grams[29].

The New Imperial Standard Pound—a pound avoirdupois—was constructed from platinum in the form of a cylinder approximately 1.35 inches in height and 1.15 inches in diameter; four copies were also made. This standard was made the legal standard of weight by the Weights and Measures Act of 1855. The description of the pound in the Act—as also in Miller's paper—refers to this pound as being equivalent to 7000 grains, the grain being the smallest subdivision of the old troy pound, which had 12 ounces, each being divided into 20 pennyweights and each pennyweight being 24 grains. Hence one troy pound was 5760 grains, or, under the new system, (5760/7000) avoirdupois pounds. Thus, the link between the two pounds was maintained, but with the primacy transferred from troy to avoirdupois.

The USA, of course, also used the pound as its unit of mass. In 1866, the US accepted the legality of both the pound and the kilogram as units, the former still derived from a standard obtained from England and the latter from a standard obtained from France. An approximate conversion factor between the two was given. In 1893, the US adopted a fixed ratio between the pound and the kilogram, the latter being based on the mass of the US national prototype provided by the BIPM. Connor reports that in 1933 the British and US pounds differed by 'nearly 19

[26] McDonald (1960) p 26.
[27] Miller (1856) p 762.
[28] Miller (1856) p 875.
[29] Connor (1987) p 293.

parts per hundred million'[30]. It was not until 1963 that the UK followed suit and defined its pound in terms of the kilogram. The Weights and Measures Act of that year also made it clear that these were standards of *mass*, not weight.

8.4 The metre convention, the BIPM and the international prototype of the kilogram

W H Miller, mentioned earlier, is best known for his foundational treatise on crystallography, published in 1839; crystallographers still describe crystal structure using 'Miller indices'. He was a member of the International Metre Commission from 1870–72. The Metre Commission eventually led to the Metre Convention, which met in 1875. This conference established the CGPM (*Conférence générale des poids et mesures*) the CIPM (*Comité international des poids et mesures*) and the BIPM.

The Metre Convention discussed new standards for the metre and kilogram and at the first meeting of the CGPM in 1889 the International Prototype Kilogram (IPK) was adopted. This was a cylinder, made by a British company—Johnson Matthey & Co.—and constructed of pure platinum alloyed with 10% pure iridium, designed so that its mass should equal that of the Kilogram of the Archives. There was no longer a reference to the density of water. The volume of the IPK had been determined by hydrostatic weighing and, after correcting for air buoyancy, it was found to be within ± 10 μg of the mass of the Kilogram of the Archives. An important feature of the new kilogram was that it had many copies, including four 'official copies' (later increased to six). This followed the decision to create multiple copies of the International Metre; it was the fact that the IPK was one of many identical copies that ultimately made it different from the Kilogram of the Archives.

Copies of the IPK were distributed to member states of the Metre Convention. Quinn gives a full account of the history leading to the Metre Convention and what has followed[31].

8.5 Relative stability of national and international prototypes

J E Sears, the Superintendent of the Metrology Department of the National Physical Laboratory, announced in 1937 that the standard British pound, made of platinum sponge, was 0.2 ppm lighter than 50 years earlier, compared to kilogram standards made of platinum–iridium alloy. He suggested comparing the Kilogram of the Archives (which was also made of platinum sponge but of an earlier vintage) with the International Prototype Kilogram.

This comparison was done in 1939. The KA was found to be 0.4 ppm lighter than 55 years earlier. Extrapolating back, it would have lost 0.6 mg between 1799 and its use in 1883. Since the latter occasion was the calibration of the IPK, this means the new standard was not the same as the original KA, linked in turn to the density of

[30] Connor (1987) p 293.
[31] Quinn (2012).

water. In terms of the best balance available in 1799, the difference was negligible but could now be detected thanks to improved technology.

The Metre Convention included a stipulation that copies of the IPK should be periodically compared with it. Three *periodic verifications* have been performed: the first between 1899 and 1911 'without recourse to the international prototype'[32]. The second was started in 1939 (which explains the timing of Sears' request) but was interrupted by the Second World War. It was re-started in 1946 and completed in 1953. The third verification took place between 1988 and 1992. Following the decision in 2011 to redefine the kilogram, a further calibration exercise was carried out in 2013 and 2014 as a prelude to the redefinition itself [33].

The second verification compared four official copies with the IPK. Of these, three had increased in mass (relative to the IPK) by around 30–40 μg while one had decreased by 30 μg. Since only relative comparisons can be made, it is not possible to say whether there has been an actual gain in mass of one standard or an actual loss in mass of the other. Girard mentions two factors which can cause an actual change in mass: contamination (which clearly increases the mass, but may be removable by cleaning) and wear during use (which will clearly decrease it)[34]. However, since the IPK *defines* the kilogram, any discrepancy with a copy must be associated with the copy, whatever the actual cause.

Two more official copies were first calibrated in 1946 and all six were then compared with the IPK in the third verification.

It was understood from the start (1882–89) that the prototype and its copies would need to be cleaned before comparison, to remove dust and other deposits. However, after 1889 this was discontinued. During the second verification, the question of cleaning was reviewed and found to be necessary. Initially the standards were dusted or cleaned with ethanol and then petrol, but a steam wash was added in order to remove solvent residues. This procedure, namely solvent cleaning (with petrol replaced by ethanol and ether) followed by steam cleaning, is known as 'cleaning and washing' and has been incorporated as a *mise en pratique* to precede all weighings[35]. During the pre-redefinition calibration exercise of 2013–14, it was found that two cleaning-and-washing cycles were necessary in order to stabilise the masses of the prototype and its copies; the third such operation 'has no significant effect'[36].

Variations in physical artefacts are clearly a worry, especially in an age when greater and greater precision is called for, or at least in an age when balance technology has again been significantly improved. Furthermore, concerns about the variation in the kilogram have even begun to appear in the national media, sometimes carrying contradictory messages and hence sowing confusion. In 2011, the 24th CGPM passed a resolution which acknowledged that 'although the

[32] Girard (1994) p 317.

[33] See Stock *et al* (2015).

[34] Girard (1994) p 326.

[35] Girard (1990).

[36] Stock *et al* (2015) p 313.

international prototype has served science and technology well since it was sanctioned by the CGPM at its first meeting in 1889, it has a number of important limitations, one of the most significant being that its mass is not explicitly linked to an invariant of nature and in consequence its long-term stability is not assured'[37]. The conference accordingly announced its intention to propose a revision to the SI system, which would involve supplying the missing link 'to an invariant of nature'—in this case, Planck's constant. The following meeting (the 25th), held in 2014, noted that progress had been made on this and set a timetable for the new unit definition to be adopted at the 26th meeting, due in 2018. As well as the unit of mass, those of electric current, absolute temperature and 'amount of substance' are due to be linked to constants of nature at the same meeting. There will then be no units defined by physical standards. In fact, it was already reported to the CGPM in 1960 that the kilogram, being an artefact, was the principal weakness of the newly announced International System of Units (SI); this situation would need to be remedied 'sooner or later'[38].

8.6 The new definition of the kilogram

The idea of defining units in terms of physical constants could be said to go back at least as far as the original metric system in 1791, if the density of water and the circumference of the earth can be regarded as physical constants. Nowadays, however, we reserve the term 'physical constant' for entities which are not specific to particular bodies or even particular substances. The first attempt to define units in terms of the *modern* idea of a physical constant was that of George Johnstone Stoney, who in 1881 devised the Stoney unit system, consisting of the speed of light, the Newtonian gravitational constant and the electron charge as units of velocity, gravitation[39] and electric charge; though we must immediately apply the caveat that what Stoney postulated was a fundamental charge, based on results from electrolysis, rather than a particle which was not to be discovered for another 16 years[40]. The Stoney units were not necessarily intended to be *practical* units and nor were the Planck units, which appeared a few years later. They were motivated by the desire for our units to be independent of arbitrary artefacts; it is not clear whether reliability and the wear and tear associated with artefacts, were also factors.

In 1983 the CGPM redefined the metre in terms of the speed of light, giving the latter quantity an exact value of 299 792 458 metres per second. The difference between this and the Stoney units is mainly one of size: the metre is effectively 1/299 792 458 of the distance travelled by light in a second. A further distinction lies in the fact that, under the CGPM definition, the speed of light itself is not used as a unit (which would require a change of basis to one that included velocity) but is linked to a unit. Other redefinitions followed and paved the way for the 2018 proposals. It is

[37] CGPM (2011).

[38] CGPM (1960).

[39] Stoney called his unit of gravitation 'the Newton'. See O'Hara (1975) p 276 note 20. However, the concept of 'gravitation' as a base quantity is philosophically somewhat problematic; see Grozier (2017).

[40] O'Hara (1975) p 269.

often forgotten that arguments for linking units to constants had already been made prior to the adoption in 1948 of the ampere as a base unit. Since then, the ampere has been defined by fixing the value of μ0, the permeability of free space (listed in the CODATA values of physical constants), equal to 10^{-7} N/A^2. The ampere definition, which had been debated throughout the previous half century, set a precedent which has since been followed.

The link between Planck's constant and the kilogram can be realised by two entirely separate routes, which gives it greater robustness. One will use the Kibble balance (see chapter 9) and the other—the *silicon x-ray crystal density method*—will involve making precise measurements of silicon spheres, on both macroscopic and atomic distance scales.

8.7 Realisation of the kilogram using the silicon x-ray crystal density method: Si→SI

Modern technology enables us both to manufacture pure, spherical crystals of silicon and to characterise them with great precision. If the crystal structure and the lattice parameter (separation of adjacent atoms in the crystal lattice) are known, the volume of the sphere can be compared with the volume associated with a single atom, to yield the number of atoms N in the crystal.

The mass of each atom is equal to the product of the relative atomic mass of the material in question and the atomic mass constant, m_u. Hence

$$m = NA_r(\mathrm{Si})m_u$$

where m is the mass of the crystal and $A_r(\mathrm{Si})$ is the atomic mass of the silicon atom relative to 1/12 the atomic mass of carbon 12.

The ratio m_u/h is a constant of nature which has been determined by various means, including the atomic recoil method, which measures the recoil velocity of an atom when it absorbs a photon of known frequency. Inserting it into the above equation yields

$$m = hNA_r(\mathrm{Si})\left(\frac{m_u}{h}\right)$$

Since the proposed modifications to the SI system include *defining* the value of Planck's constant when expressed in the unit kg m^2 s^{-1} and the ratio m_u/h is measured in m^{-2} s, the other quantities in the formula being pure numbers, this will deliver a mass in kilograms for the crystal, which can then be used as a secondary mass standard. The only remaining problem is to ascertain $A_r(\mathrm{Si})$. The crystal will contain a mixture of the three isotopes of silicon and the relative abundances must be measured. Relative atomic abundance, like relative atomic mass, is a pure number and therefore independent of any system of units.

Following the redefinition of the kilogram, the exact value of Planck's constant specified in it will allow the realisation of the kilogram at the atomic scale in terms of the ratio h/m_u and at the macroscopic scale by direct comparison with the mass of such a well-characterised silicon sphere (the crystallographers Abbé Haüy and W H

Miller could only have approved), or by means of a watt balance. Any suitable method can be used. The *x-ray crystal density* method described above will provide a link between the microscopic and the macroscopic and an independent route from *h* to mass standards[41].

8.8 Conclusion

We have seen how, as techniques and precision improved with time, metrologists in both France and Britain have always at least *attempted* to ground standards on 'natural' quantities. With the new SI definitions, it might appear that we have at last found a way to preserve those 'standards' indefinitely, albeit in the somewhat abstract context of fundamental constants. Nevertheless, we should bear in mind the words of Charles Édouard Guillaume, BIPM Director and Nobel laureate, who prophesied in 1920, less than a century before the demise of the IPK, that 'the unit of mass is assured to about 1×10^{-8} for more than 10 000 years'—although he did add that 'no doubt, well before it ends, work carried out by metrologists in future centuries will have led to even more flawless solutions'[42].

References

CGPM 1960 *Proceedings* 25 http://www.bipm.org/en/worldwide-metrology/cgpm/resolutions.html

CGPM 2011 http://www.bipm.org/en/CGPM/db/24/1/

Chisholm H W 1873 On the science of weighing and measuring and the standards of weight and measure *Nature* **VIII** 388

Cladé P, Biraben F, Julien L, Nez F and Guellati-Khelifa S 2016 Precise determination of the ratio h/m_u: a way to link microscopic mass to the new kilogram *Metrologia* **53** A75

Connor R D 1987 *The Weights and Measures of England* (London: HMSO)

Davis R S, Barat P and Stock M 2016 A brief history of the unit of mass: Continuity of successive definitions of the kilogram *Metrologia* **53** A12

Gillispie C 2004 *Science and Polity in France: The Revolutionary and Napoleonic Years* (Princeton, NJ: Princeton University Press)

Girard G 1990 The procedure for cleaning and washing platinum-iridum kilogram prototypes used at the BIPM (BIPM)

Girard G 1994 The third periodic verification of national prototypes *Metrologia* **31** 317–36

Grozier J 2017 Fundamental constants, unit-invariance and objectivity *Studies in the History and Philosophy of Science* (under review)

Guillaume C 1927 La Création du Bureau international des poids et mesures et son œuvre (BIPM)

Kershaw M 2012 The '*nec plus ultra*' of precision measurement: Geodesy and the forgotten purpose of the Metre Convention *Stud. Hist. Phil. Sci.* **43** 563–76

McDonald D 1960 *A History of Platinum* (Johnson Matthey & Co.)

McGreevy T 1995 *The Basis of Measurement, volume 1* (Picton)

[41] Cladé *et al* (2016) p A76.

[42] Guillaume (1927).

Miller W H 1856 On the construction of the new imperial standard pound and its copies of platinum; and on the comparison of the imperial standard pound with the kilogramme des archives *Philos. Trans. R. Soc. London* **146** 753–946

O'Hara J 1975 George Johnstone Stoney FRS and the concept of the electron *Not. Records R. Soc. London* **29** 265–76

Quinn T 2012 *From Artefacts to Atoms* (Oxford: Oxford University Press)

Riordan S 2013 The making of the kilogram, 1789–1799 *PhD thesis* (Stanford University)

Riordan S 2015 The objectivity of scientific measures *Stud. Hist. Phil. Sci.* **50** 38–47

Stock M, Barat P, Davis R, Picard A and Milton M 2015 Calibration campaign against the international prototype of the kilogram in anticipation of the redefinition of the kilogram part 1: comparison of the international prototype with its official copies *Metrologia* **52** 310–6

IOP Publishing

Precise Dimensions
A history of units from 1791-2018
Malcolm Cooper and Jim Grozier

Chapter 9

Mass from energy—a unit for a quantum world

Bryan Kibble

This is a transcript of the talk given by Bryan Kibble at the National Physical Laboratory, UK, and although we have made a few changes to make the account clearer, the reader should bear in mind that this is essentially a verbatim transcript.

I thought that we're all used to reading papers of work which has been done which present a beautifully, logical clear picture of the beginning of the work and solving some problems, going through to results and a conclusion, lovely, clean. Us in the trade know that things are not really like that so I thought you might be interested to know what the real story of the watt balance is—it's quite interesting.

Let's begin with this current balance—this venerable old instrument at the NPL (figure 9.1). Various countries have current balances and their purpose in life was to get at the ampere. Current balances go back to before the beginning of the 20th century—the Ampere definition came into being to suit the current balances—not the other way round—comparatively recently. So, it's served well for many decades but it had some problems. It consists of coils hung from a balance which are suspended inside fixed coils and you weigh the force on the coils inside. When all the coils are fed with a current of a nominal one ampere, that produces a few miserable grams of force—kind of difficult to weigh with any great accuracy. Worse than that, unbelievably, there are 50 watts of heat produced in the balance case by the current going through these coils, and this, of course, sets up convection currents. Again unbelievably, these convection currents are stable enough that you can work with the thing for about a couple of hours and then they go chaotic and unstable and you pack up for the day and go home and wait for it to cool down again over night. A plate was put across between the coils and balance beam—to try and alleviate that problem—but, well, it did a bit, but not greatly—but there are other problems.

In order to calculate what the force should be and realise the two-parallel-wires definition, you just wind the parallel wires up into coils. To do that you have to

Figure 9.1. Current balance. Courtesy of NPL. Photograph by Ian Robinson.

know the position of every single point on every single wire on this complicated coil system meaning hundreds if not thousands of length measurements, done to the utmost possible precision. And that is not all because the wires are straight wires deformed into circles, winding around; consequently they are strained across their cross section. Strain alters the resistance of metals and so you do not know where the current is flowing, as it has a varying density across the cross section of the wires, and this you have to take a guess at. And one or two other things as well and the net result of this is that the current balance had one sigma error of about 20 parts per million—and that was fine when electrical measurements were not that precise.

But by the beginning of the 1970s people had developed digital voltmeters which they were selling between countries and they had a resolution and a stability of a part per million. The current balance could only manage 20 parts per million. Worse than that, every national laboratory worth its salt had a version of this balance and they all quite properly thought that theirs was giving exactly the right answer. The result of that was there was an English volt, an American volt and of course a very divergent French volt: 9 parts per million divergent I think—so the DVM manufacturers began to get rather cross about all this and clearly something had to be done.

Now, to deviate a little bit. I at that time was also hanging a coil up from a balance—my coil was in a strong magnetic field of a permanent magnet and its purpose in life was to determine the gyromagnetic ratio of the proton—that is the rate that protons precess in a magnetic field, which you measure by an NMR technique in terms of the strength of the field which you measure with a coil of known width. This was a rectangular coil hung up from a balance and we weighed the force. So I was busy with that, and the work was not going all that well, but I'd learnt of some precise coils which had been made by a Dr Briggs in the Australian National Laboratory, and unfortunately the management of the Australian laboratory decided that, since he was retiring, his work should stop. So I thought: 'Aha! if I can take over his coils—which were a beautiful bit of Australian workmanship—glass with a strip of silver conductor, rectangular section round the edge, and you could measure the width between the spaces of the rectangular conductors—I could get on a bit faster'. So I wrote and asked and the Australian laboratory very kindly said well, yes, you can have them, of course. So they packed up these fragile glass coils and sent them across to me, and Dr Briggs came with them. I reckon he fancied six months in the old country—fine—and we set about it, and it went pretty well from then on—we weighed the coils and so on.

But one day there was a bit of an accident in that this fragile glass coil suspended in the jaws of this enormous straight permanent magnet—imagine it if you will—suddenly decided to take it upon itself to flip like a galvanometer coil and crash against the pole face of the magnet. The damage wasn't too great and it was still actually useable but I said some rude words. Dr Briggs was very kind—he said 'well, don't get cross with the coil—it's really not its fault—its energy is just the product of the current in it times the magnetic flux threading it, and all it was doing is trying to minimise its energy by rotating transverse in line with the flux, so the flux threading it was smaller'.

Well, this kind of sounded very interesting to me because this theorem—I don't know whether any of you know it, that the energy of a current—a circulating current in a circulating magnetic flux—is just the product—(and they *must* link)—is just the product of the flux times the current, and is *topological*—it doesn't matter where it is, how they thread, what shape the coil is, it doesn't matter what anything is, the energy is just the energy of the linkage. Well that was fascinating, so I stored it in my head filed under 'quite interesting' and went on playing, and got a result for the gyromagnetic ratio of the proton; and then, this current balance is in trouble. Dr Vigoureux[1] was about to retire and I could sort of see that this machine would be coming my way to improve, by a factor of at least 20, and that I would have to perform all these thousands of measurements, and—try as I might—I did not like this idea one bit. So, in desperation I started to think along the lines of 'Well, OK, there's a force due to the current and so on and it depends on the size of the coils. Now, suppose we move one coil inside the other—that will generate a voltage and the voltage will also have something to do with the geometry of the coils, won't it?'. This idea went round in my head for a little while getting not very far when I suddenly remembered Dr Briggs and his incredible threading theorem— $E = I\Phi$ where I is the current in some loop or other, Φ is the total amount of flux threading it and some elementary little bit of calculus somehow entered my head. Ok, that's a

[1] Paul Vigoureux—metrologist at NPL. http://www.npl.co.uk/people/paul-vigoureux

force, that is dE/dx and so there we are—let's weigh the thing— I was used, you remember, to weighing coils in fields. *Mg* the force is $I\ \partial\Phi/\partial x$. OK, what about the voltage? The voltage, as we all know, is $\partial\Phi/\partial t$. One day it just suddenly entered my head—hang on, that's $\partial\Phi/\partial x$ times dx/dt isn't it— yes, OK dx/dt is velocity and $\partial\Phi/\partial x$—where have I heard that before? Oh there—what happens if I eliminate it between the two equations? All this took about 2 seconds—in 5 metres walk from the library getting to 6 metres from the library—hence this—and the watt balance was born.

You can measure *I* by the volts developed across a resistance *R* and then you have an actually usable equation. By this time—the early 70s—the calculable capacitor[2] had been invented and so on, resistance was in good shape so you could get to current from the voltage by Ohm's Law, so that was no problem. So here we have a replacement for the current balance. Now, sometimes in life—mostly you're unlucky. On the way here this morning—we have two independent bus services running roughly every ten minutes, and I arrived at the other side of the road crossing, the bus stop on the other side, just in time to see one kind of bus immediately behind the other kind of bus, just receding into the distance—that's life—bad luck is everywhere. But this turned out to be *stupendous good* luck.

It took us quite a while to realise how good it was, because there are several advantages over the current balance—no coil size, shape or current path measurements needed—remember it doesn't matter how the toroids thread one another so a lot of that goes out.

Now to do the experiment we're going to split it into two halves. First we're going to weigh to determine the force with a current and secondly we're going to move the coil and measure the voltage. Splitting the measurements in two halves like that means that you're not talking about *actual* work—if you were you would worry about a friction in the bearings, air resistance and Joule heating in the coil and all sorts of stuff like that but it's *virtual* work. When you're weighing the coil—voltage doesn't come into it—the coil isn't moving and when you're moving the coil there's, in principle, no force on it so no actual work is done in bringing about that equation that I pointed out. And, all you need, in principle, is one measurement of the velocity in order to satisfy that equation. It has to be vertical, the velocity, because by definition the force from weighing is vertical.

But, but, (there is always a but, isn't there?)—we obviously have to eliminate any velocity which might accidentally occur in the other five degrees of freedom. As well as moving up, the coil might well move a bit sideways—it might rotate about the free axis a little bit and so on—all of those would produce a little extra voltage, wouldn't they? and that would cause an error. So, it's quite vital that we align the coil to move vertically, and fortunately, since we're talking about vectors, we don't have to do this all that accurately, so it's not a big limitation but—there's a but.

[2] The calculable capacitor allows the calculation of the capacitance of a special type of capacitor directly from a single dimensional measurement which can be made traceable to the SI unit of length. (BIPM) See chapter 3 figure 9.10.

$$E = I\Phi$$

(1) $Mg = F = I(\partial\Phi/\partial x)$

(2) $V = (\partial\Phi/\partial t) = (\partial\Phi/\partial x).(dx/dt) = (\partial\Phi/\partial x).u$

Hence $IV = Mgu$

Measure I by volts V' across R,

$$VV'/R = Mgu$$

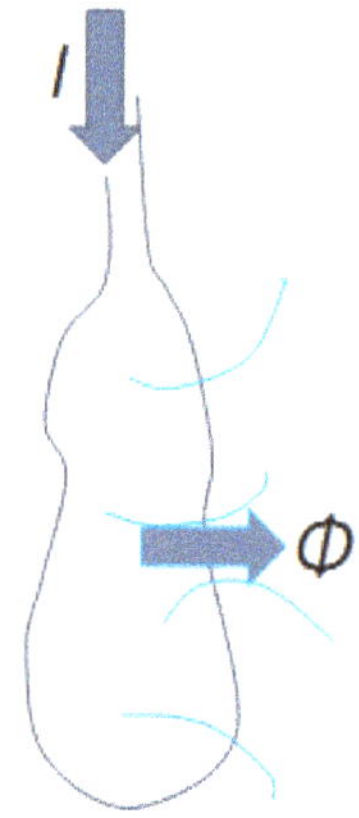

Now this again is a bit of stupendous good luck—I think it is anyway: the things you've got to measure are all favourable magnitudes to measure to—well, these days, getting on for a part in a billion. The current is about 10 mA so it will produce 1 V when it flows through a 100 Ω resistor; the velocity—well if we have an interferometer to measure it—that's typically 5000 optical fringes a second and 10^{-5} of an optical fringe is do-able—we have to be careful but it's do-able—again a part in 10^9 sort of stuff. *g*—that's depended as the years went by on the development of a commercial gravimeter, but now it claims accuracies of about 3 parts in a billion so that's alright. And the kilogram—we have to weigh to a few micrograms to get a few parts in 10^9 but there's a curious thing about balances, however big and heavy they are—ours is massive—they seem to have a resolution of micrograms, if you're careful. That was true of the old current balance too, before convection currents got going. So, the measurements are all of a suitable size—I don't know if that's elementary physics or stupendous good luck, depending on which way you look at it.

Here's our first watt balance.

It works in air—now that was a limitation because the refractive index of air has to be measured by somebody and used to correct the distance versus velocity measurement and also its buoyancy on the kilogram and both of these are a few hundred parts per million. So to get the accuracy we wanted out of this, you have to measure the density and refractive index of air very carefully. This was done[3] by the people we had to measure the barometric pressure, the temperature and all the things that go into a formula for it. Anyway, we did that—this is my great big permanent magnet (figure 9.2)—it's several tons; its pole faces are—excuse the imperial units—about a foot in diameter. The coil is a rectangular coil—it is not a precision coil—it does not need to be—it could be any old shape in principle and the weighing is by substitution. The kilogram is on the right side and it's done in this manner: you first of all lower a 500 g mass—which doesn't have to be precise—on

[3] The equipment was on loan but we did the measurements.

Figure 9.2. NPL Watt balance. Courtesy of NPL. Copyright Ray Smith.

the right side of the balance and oppose it with a current pulling the coil down on the left side—you observe that current. You then lower the actual kilogram you want to calibrate or use on this side and reverse the current which produces an equal push up and the balance is in equilibrium again. That makes the current change symmetrical which is important for avoiding hysteresis problems in the magnet—and other things—so that's a useful thing to do. So that's how it works and that's it.

Now, a little divergent story—at the beginning after this light on the road to Damascus—the phenomenon of inventing the formula for the watt balance—the first thing I did was write a feasibility study in order to screw some money out of the NPL management to actually do it. And there was a conference at that time and Bob Cutkosky from NIST[4] came to that conference, and he is my ultimate hero as a metrologist, he's fantastic and very, very bright. So I showed him this feasibility study and said 'do you think this'll work Bob?' He took it away with him overnight, appeared the next morning and said 'yeah, I think it might' and I knew then that I was on to a sure fire winner. But anyway, he went back to NIST and the next thing I heard is that he'd rigged up an apparatus to sort of try it out—a crude apparatus—and the next thing after that I heard was that they had got a watt balance of their own—which many of you will be familiar with.

Like many things in the States it is BIG, it is HUGE—it is two storeys high from the bottom to the top—which keeps you in trim going up and down the stairs. In detail it's very different but the principle is exactly the same. The coils are horizontal coils, and the force is provided by a superconducting solenoid—which is a right gas

[4] National Institute of Standards and Technology (USA).

guzzler of liquid helium—and the balance is a wheel rather than a beam and ribbons of parallel fine wires come off the polished circumference of this wheel and act like a balance arm. They did that to make the movement purely vertical—which it always is if the coils and wheels are exact enough. And that worked.

Now we had our first success, (remember the current balance was 20 parts per million) this, more or less straightaway, was 100 times better—bang—just like that. So, with some careful work to make sure that we had no wrong answers, problems, systematic errors—the usual sort of thing—takes years unfortunately. By 1990 the BIPM was able to recommend that the electrical units now could be represented (a very good choice of words) by fixed values of the Josephson constant (the voltage) and the (quantum Hall) von Klitzing constant (the resistance). And the absolute values, if you like, were not very different from these purely defined values and the world's mostly content with the defined values. So that was a great victory for us and we had a little celebratory party at our house on 1 January 1990 when we stayed up to midnight in order to see the lights of the house go dim because the voltage unit had changed by a part per million. And there was a leap second, and whether we could see that on the newly developed digital radio-controlled clock—which as some of you may know had been developed at NPL.

Unfortunately, by midnight we had consumed the equivalent of too many alcohol thermometers—but never mind. Anyway, that was a success—oh—and one extra, enormous bit of good luck was right at the beginning of this work, I was joined by Ian Robinson and the good luck was that he possessed all the things that I did not know. You'll have gathered by now that I'm pretty computer-illiterate and he was a whizz kid with it, and also with electronics, which we needed for low noise pre-amplifiers and that kind of thing. He possessed, in short, all the skills that I did not possess to make the darn thing actually successful and work. So, it worked and was successful and then I can remember quite distinctly sitting in the lab with Ian after the 1st of January, 1990 and saying 'that was alright—now what do we do?'

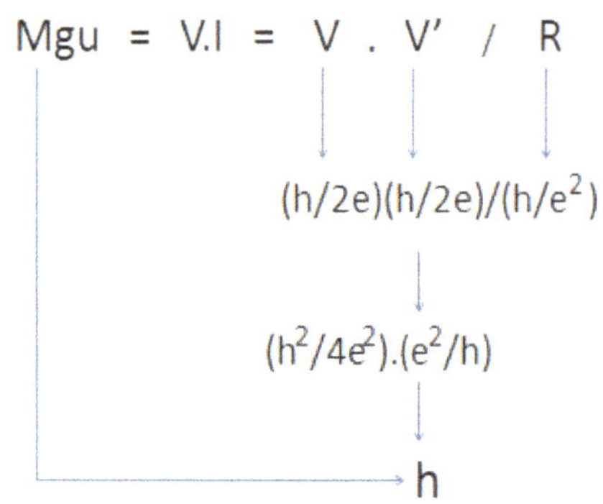

Well, we sat and talked for a minute but by that time we'd kind of realised that the *V* times I can be written as *V* generated by the coil times *V'* the voltage generated by the current going through the resistance *R*. The *V*s were linked to the Josephson constant—by that time we had linked our apparatus up to a Josephson array and *R*

could be measured by cryogenic means in terms of the von Klitzing constant. So that's how our measurements were derived. Now if you just work that out (referring to figure 9.4) you find that the *e*s cancel provided it's the same *e* in both cases—which was a little bit of a question but the theoreticians assured us that they probably were—you have a direct link—you're left with *h*, some integers, some frequencies and that's all. You have to measure *g* and the velocity *u* of course but you have a direct link between *m* and *h*. Mr Einstein smiled in his grave because that's mass related to energy—lovely.

So if someone would kindly tell us what Planck's constant was, numerically, we could tell them what their mass was. We could generate a newly defined kilogram—OK right let's get on with it—out of our armchairs, roll up our sleeves—what do we need to do? Well the old apparatus worked in air and achieved an accuracy of about 200 parts in a billion. We needed (the mass people at BIPM told us we really needed) to make that target about 20 parts in a billion. So that's an increase of 10 times in accuracy—well, that's alright, that's easy peasy. So, the first thing to do is put the apparatus in a vacuum—so we did and there's Ian Robinson (see figure 9.3). (At this point Bryan was given a time reminder of 10 mins to go but says jokingly 'but I'm only on page 1! OK —we'll get there!'.)

There are changes in geometry—gone is the rectangular coil, it is now a toroidal coil—in the lower vacuum chamber—and the magnet is essentially a loudspeaker

Figure 9.3. The Watt balance (now re-named 'The Kibble balance'). Courtesy of NPL. Photograph by Bryan Kibble.

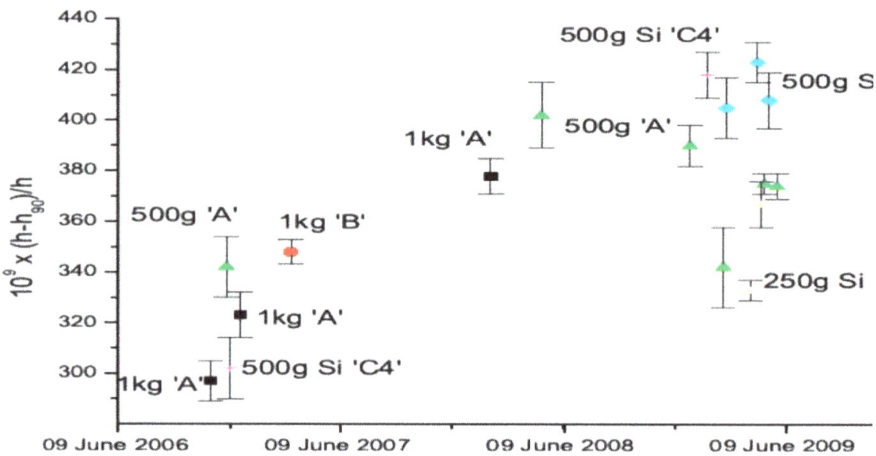

Figure 9.4.

magnet sitting on its back. But it's the same old beam, it weighs 44 kg—but it works so we got on with it—and that's where all the trouble started!

The years went by and various results were obtained, (see figure 9.4) and the scatter is a bit horrendous and—observe the time scale—we're now talking of a substantial period of 3 years.

3 years! It was obvious to us that we were not aligning the apparatus well enough, and each time we re-aligned it we got a different alignment and so a different result. So Ian smothered the apparatus with auto collimators and mirrors (to look at small rotations) and non-contacting distance measuring gizmos (to look at sideways movement) and went on. Now, I must just say that this looks terrible. By this time—I'd retired and left Ian in the lurch—I was a really rotten metrologist—but there are extenuating circumstances. One is that we had other things given—other things to do during this time. The other thing is that all sorts of little side investigations like the hysteresis of the knife edges and little things like that were done in this interval and dealt with, problems solved and so on—all that was good[5].

Also in this time the NPL management decided that really ours looked like a rough old apparatus and we really should have a gold plated one and they set up a committee to design the next version of the watt balance. You know the old saying that a camel was designed by a committee—so Ian had to spend about 3 years of his time trying to ward off the more silly ideas generated by it—anyway! We finally arrived at the situation where the NPL management really excelled itself and sold the whole apparatus lock stock and barrel and indeed some of Ian's time, to Canada. You will not believe this—but it's absolutely true—a date was set for the apparatus to be packed up in a container to be shipped to Canada—absolutely immovable date. Two weeks before that date Ian suddenly came to the conclusion that because

[5] The important issues were to do with rotations but it took quite a while to find out and fix the issue.

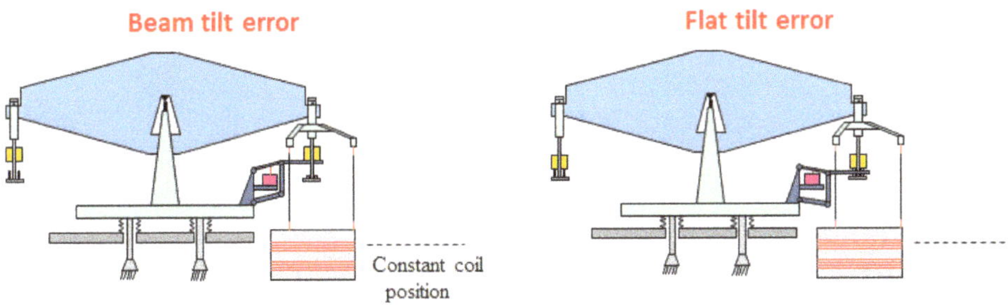

Figure 9.5.

of all these measurements with auto-collimators we'd made—the alignment really was good enough—it must be something else.

And he thought of the something else! This is really stupid—we had decided in the beginning that we didn't need a pointer for the balance—as most ordinary balances do—because we were measuring the coil's position with an interferometer and, in fact servo-ing the coils to a constant position—and that was a monitor of the beam angle and we didn't need a beam pointer. Wrong!

Keep an eye on this 500 g counterweight here (figure 9.5)—it's shown in the lower position—it's raised—remember this is servo'd to a constant position. These rods stretch and the beam is at a different angle and the balance was not adjusted to be asymmetrical—it had a restoring force due to its tilt. There's one error and rather quickly there's another one—and keep an eye on the kilogram mass this time—when that happens the base of the balance tilts and the central knife of the balance tilts and so its point of contact changes—another error. The two errors that Ian found in these two weeks left him, do sort of cancel but not completely and he had no time to investigate their magnitude and make a correction and anyway it obviously changed over the years—that's why there's scatter. We had designed in some weights right at the beginning to make it balance the theory of it, and never used them—how stupid was that?

We went to NRC[6] with this knowledge and imparted it to them and they made some simple modifications and the weighing went from what you see on the left—immediately to what you see on the right (see figure 9.6)—much, much, better, much more consistent.

[6] National Research Council of Canada.

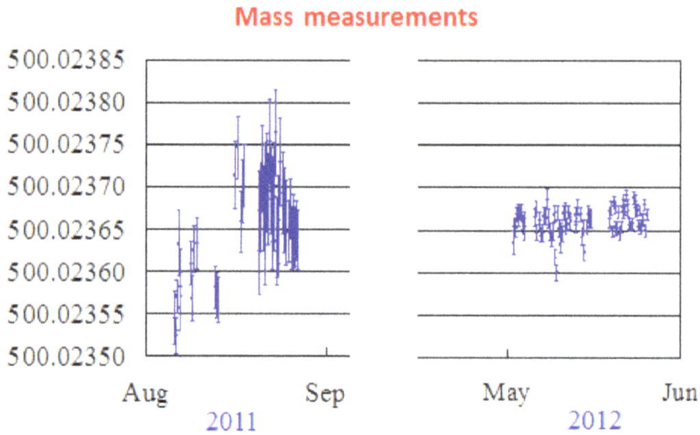

Figure 9.6. Courtesy of the National Research Council Canada.

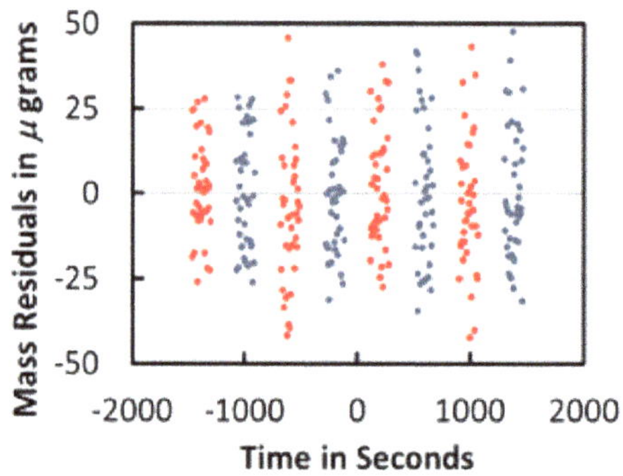

Figure 9.7. Courtesy of the National Research Council Canada.

Here's a set of weighing results (see figure 9.7)—on the left is the scale in parts per billion essentially and quite clearly an average of a few parts per billion, no problem. Velocity was not so good.

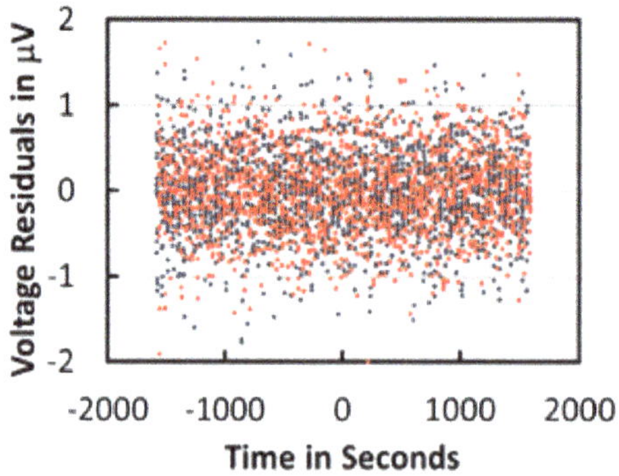

Figure 9.8. Courtesy of the National Research Council Canada.

The voltage generated divided by velocity is plotted here (see figure 9.8)—this is a succession of up and down movements—lots and lots of them going every ten seconds or so and now the scale, the one sigma deviation of it—is a substantial part of a part per million—remember we are aiming at a few parts in a billion, so that means you have to collect an awful number of points to average them. And clearly an elephant sized systematic error could be hiding in that long grass. The Canadians saw that and quite properly were very dissatisfied with it—here again it is on the left—(see figure 9.9) and spent a couple of years trying to track down where this additional noise was getting into the system. After a couple of years they got the idea that it was something to do with the interferometry—that the wave front going through the interferometer was not as flat as it could or should be.

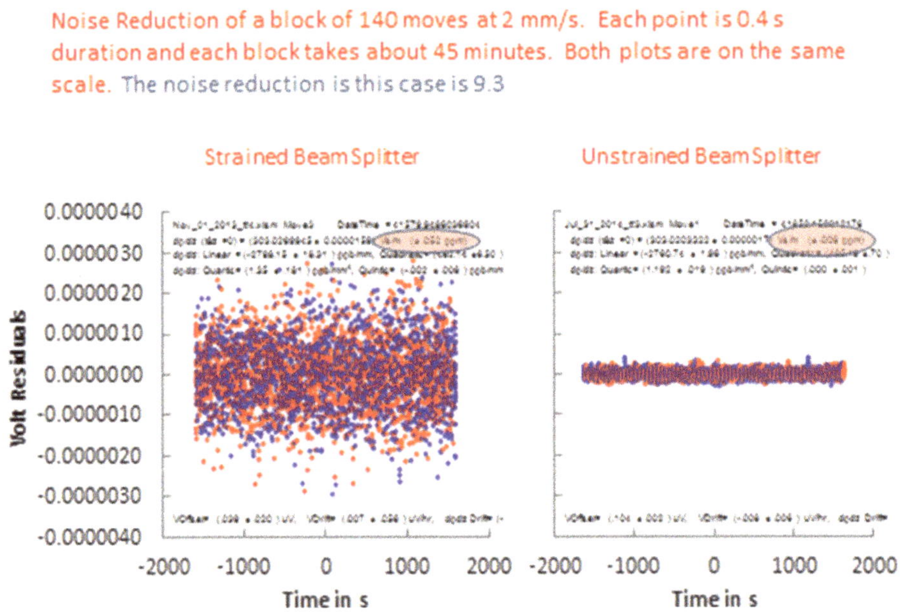

Figure 9.9. Courtesy of the National Research Council Canada.

So, again this was probably my fault in that the 45° mirror which reflected the beam up—I had put four screws in to hold it. Anyway, one of the screws needed to be loosened so that the mirror could adopt its proper flat surface. (This was really down to both of us but was not very obvious as it took two years to find and the discovery was almost by accident.) So, the smallest member of the Canadian team volunteered to crawl under the apparatus and in a few minutes had loosened the screw and the results you see on the right. (At this point the chairman says ‘Bryan, you’re going into extra time’. Bryan Kibble quickly replies—‘do I get paid for overtime?’.)

This means that it's a tenth of the peak to peak noise—roughly—it means you need one hundredth only of the data points to get the same resolution. So they could investigate a lot of systematic errors.

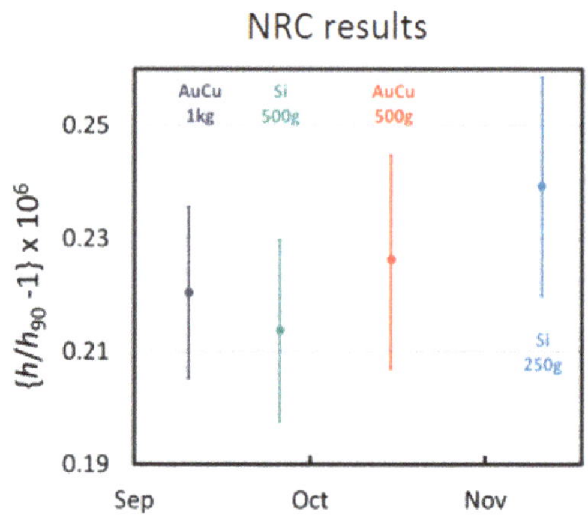

Figure 9.10. Courtesy of the National Research Council Canada.

Let's pass on the NRC result (see figure 9.10)—it's important you check that you get the same result for several values of the mass because of possible non-linearity of the interaction of the current and the magnetic flux. The scale you will notice is graduated 20 parts per billion apart. I think you would agree that about 20 parts per billion could be ascribed to that—that's better—not the only results they have, of course. The result was the value of h—is the important one—19 parts per billion—and remember that the specification of the mass people—to protect their kilogram—should be a result of 20 parts per billion or less—so that's good.

The work on silicon atoms which Richard in the last talk (chapter 8) mentioned came out with the result about the same uncertainty; and what is incredible (if you're a cynic like me) or wonderful (if you're not) they agree within that uncertainty. The NIST group struggled on with their monster over the years and they obtained 35 parts per billion error, and within that error they agreed with this. So we would maintain that as of this present time all is well with re-defining the kilogram.

Here is something where our friend Murphy—he of the law of maximum costliness—you can visualise rolling about the floor howling with laughter because the Canadians who had the apparatus found experimentally and Ian who did not have the apparatus found theoretically—more or less independently, and more or less at the same time that *alignment doesn't matter*. You have to do the sums properly remembering that it's a vector situation—all these misalignments—little bits of velocity and rotations add up to modify the current balance equation—but it's still exact.

Extinct?

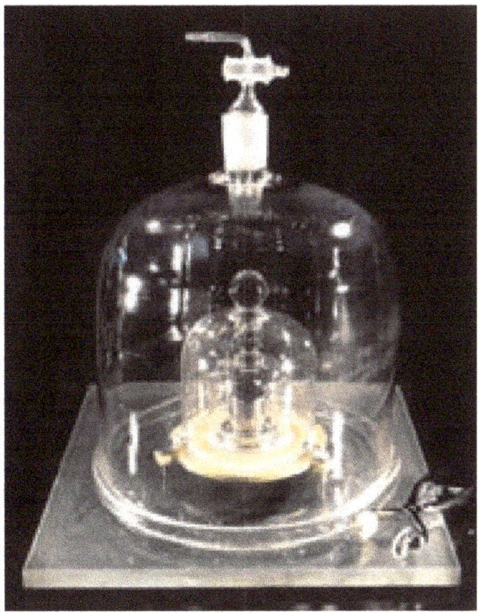

The lesson from that is to beware of the obvious. And final slide (right: courtesy of Bureau international des poids et mesures) is the grand *k* in the museum—well, no it will still be used but I have to justify the title of my talk and the defining *h* will be for the purposes of the people who do measure fundamental constants to ensure their consistency—or not, if there's some new physics.

Thank you very much!

www.ingramcontent.com/pod-product-compliance
Ingram Content Group UK Ltd.
Pitfield, Milton Keynes, MK11 3LW, UK
UKHW051852100926
8556IPUK00005B/582

9 780750 314855